Rockbursts: prediction and control

Papers presented at a symposium, organized by the Institution of Mining and Metallurgy in association with the Institution of Mining Engineers, and held in London on 20 October, 1983

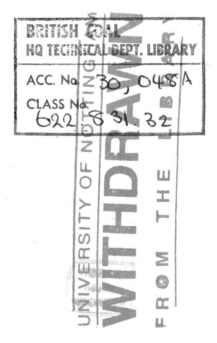
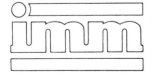

The Institution of
Mining and Metallurgy

Published at the office of

The Institution of Mining and Metallurgy
44 Portland Place, London W1, England

© The Institution of Mining and Metallurgy 1983   ISBN 0 900488 67 0   UDC 622.831.32

Printed in England by Barnes Design + Print Group.

Foreword

Rockbursts occur when overstressing of rocks in the vicinity of mined excavations results in the rapid collapse of the rock and consequent sudden release of stored strain energy. The conditions that produce rockbursts are usually a combination of large stress changes in deep mines and are associated with isolated zones of strong rock. Although these general conditions are known, it is not yet possible to predict the occurrence of individual rockbursts, but the frequency and magnitude of rockbursts can be controlled by careful design of mine operations and evaluated by monitoring of the seismic activity caused by rock fracture.

The aim of this symposium has been to define the existing state of knowledge on the occurrence, prediction, control and monitoring of rockbursts. The format of the symposium has presented a unique opportunity for mining engineers to update their knowledge of rockbursts. It has also provided an opportunity for practical engineers and researchers to discuss the future development of research into rockburst prediction and control. Although primarily concerned with rockburst hazards in mines, the symposium has also dealt with similar phenomena in tunnels and deep geothermal energy exploitation. The proceedings will thus be of interest to practical engineers in a variety of disciplines.

The Institution of Mining and Metallurgy and its co-sponsor, the Institution of Mining Engineers, have been fortunate in being able to attract such a distinguished group of international experts to present the results of their experience at this symposium. The papers have covered both theoretical and practical aspects of the problem and have included case histories from Canada, India, Norway, South Africa, the United Kingdom, the U.S.A. and Zambia.

This conference volume has been made available to delegates prior to the symposium. The technical discussions and the general report by Professor E.T. Brown will be published subsequently in the *Transactions* of the Institution of Mining and Metallurgy.*

The assistance of the other Organizing Committee members, Dr. Ian Farmer and K.C.G. Heath, is gratefully acknowledged, together with the very effective secretariat service provided by the Conference Officer, Penny Gill.

Laurie Richards
Chairman, Organizing Committee
August, 1983

Trans. Instn Min. Metall. (Sect. A: Min. industry), **93,** April 1984.

Organizing Committee

Dr. L.R. Richards *(Chairman)*
Dr. I.W. Farmer
K.C.G. Heath

Contents

Origin of rockbursts

N.G.W. Cook Ph.D.
Department of Materials Science and Mineral Engineering, University of California, Berkeley, California, U.S.A.

SYNOPSIS

A model of the formation of the fracture zone and of the mechanism of rockbursts is needed if the origin of rockbursts is to be understood properly and effective steps are to be taken on an informed basis to control or predict them. In this paper, an attempt is made to marry the wealth of practical observations that have been made of fracture zones in the vicinity of longwall stopes at depth with such theoretical and experimental work concerning the failure of rock as seems applicable. A plausible model for the fracture zone is given in terms of three types of fractures, namely inclined shear fractures, cleavage fractures, and vertical shear fractures. The last of these is a more likely candidate for the origin of rockbursts than are the other two. It is unlikely that this model provides either an accurate or a complete description of the genesis and morphology of the fracture zone or of the mechanism of rockbursts. Its principal purpose is to provide a common framework and focus for further analysis, experiment, and observation that will be used to refute, modify, or support this hypothesis.

INTRODUCTION

Rockbursts are violent rock failures that occur in proximity to underground excavations. In many respects rockbursts resemble earthquakes. Rockbursts have long been recognized as a major hazard when mining hard rock at depth. Such rockbursts have been reported from Canada[1], India, South Africa[2], the United States of America[3], and other geographic locations.

When excavations are made underground, significant changes in the potential and strain energies of the rock in their vicinity occur. The potential energy of the mass of rock removed from the excavation and hoisted to surface is increased by an amount equal to the product of the mass of the broken rock, the gravitational acceleration and the depth from which the rock is hoisted. Conversely, the potential energy of the intact rock around the excavations is decreased by the integral of the product of virgin tractions that existed across the surfaces of the excavation before they were mined and the displacements of these surfaces that result from mining. If the far-field stresses remain constant, the rock behaves elastically and the displacements of the surfaces of the excavations are insufficient to produce closure of the excavations, one half of the decrease in the potential energy of the rock around the excavations is stored as concentrations of elastic strain energy in the rock around the excavations and the other half must be dissipated or released. If the rock behaves inelastically or closure occurs across the excavations, less than half the decrease in potential energy is stored as strain energy and the remainder must be dissipated or released.

One explanation for the origin of rockbursts is that they are unstable releases of some of the decrease in potential energy of the rock around the excavations. Another explanation is that the changes brought about by mining merely trigger latent seismic events that derive mainly from the strain energy produced by geological differences in the state of stress. Both explanations may be correct.

Rockbursts occur in mine excavations of many different geometries. To facilitate the further evaluation of the phenomena described above, it is convenient to consider a specific geometry. The geometry that has been selected for this purpose is that of a long, thin, horizontal excavation such as may result from mining an isolated longwall stope. This is not necessarily the most adverse geometry in terms of rockbursts. Indeed, experience suggests that longwall mining may diminish the hazard of rockbursts compared with more complex excavations. However, this shape is simple to describe and analyze, it has been studied extensively from other aspects in fracture mechanics, and it approximates the geometry of many mining situations for which field data exist.

In this paper an attempt is made to marry the wealth of practical observations that have been made of the geometry of fractures in the vicinity of near isolated longwall stopes at depth in hard rock with such theoretical and experimental work concerning the failure of rock as seems applicable, to produce a model of the fracture zone. Although it is unlikely that the product of this effort provides either a complete or an accurate description of the genesis and morphology of the fracture zone, it will have been worthwhile if it encourages analyses and observations to be made that either refute or support this hypothesis. Whatever the outcome, progress will have been made towards a better and more comprehensive understanding of the formation of a fracture zone and the mechanism of rockbursts.

ANALYSIS

Consider an excavation in the form of a long, thin horizontal slit of width 2c in a rock mass at a depth below surface very much greater than the thickness of the slit. Changes in the stresses and displacements of the rock mass can be studied readily by imagining that a thin slice of rock in the plane containing the slit is replaced by an array of jacks that exert forces across this plane equal to those produced by the stresses that would exist in the rock. To generate an open slit, those jacks over the area of the open slit must be released. As these jacks are released,

convergence occurs between the roof and the floor of the slit and the rock mass does work against the jacks and stores strain energy in the rock, the sum of which is equal to the integral over the width of the slit of the product of the virgin component of the vertical stress in the rock at the depth of the slit and the convergence between the roof and the floor[4]. From the elastic solution for the displacement between the roof and the floor of such a slit, it can be shown that the work done against the jacks is given by:

$$W_R = \pi c^2 (1-\nu) \sigma_V^2 / 2G, \qquad (1)$$

where W_R = the work done by the rock mass against the jacks;

c = the half width of the slit;

ν = Poisson's ratio of the rock;

G = the modulus of rigidity of the rock,

and σ_V = the vertical stress in the rock at the depth of the slit.

According to (1) the rate at which energy is released as the width of the slit is enlarged in an elastic rock is given by:

$$\frac{\partial W_R}{\partial c} = \pi c (1-\nu) \sigma_V^2 / G. \qquad (2)$$

To maintain equilibrium, the magnitudes of the stresses in the rock around the slit must change. In particular, the magnitudes of the vertical component of the stress across the intact parts of the plane containing the slit must increase to compensate for the elimination of the stresses between the roof and the floor of the slit. As a result of this, intense stress concentrations arise in the rock adjacent to the ends of the slit. The magnitudes of these stresses may be sufficient to cause the rock in these areas to fail and produce a "fracture zone." The imaginary jacks can be used to evaluate the consequences of forming this fracture zone.

In Figure 1 are illustrated one edge of the slit, the vertical stress that would exist if the rock behaves elastically, and the stresses and displacements that may occur as a result of the formation of the fracture zone. From this Figure, the rate of energy release in the presence of a fracture zone can be evaluated. Assume that $F(x)$ describes the magnitude of the vertical stress as

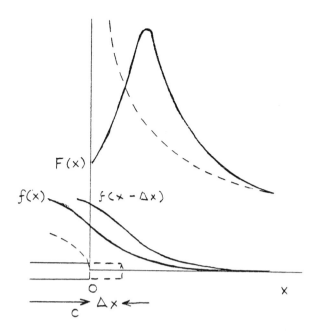

Figure 1. A sketch illustrating the vertical component of stress, F(x), in the fracture zone ahead of a slit-like excavation and the vertical components of displacement, f(x) in this zone and behind the end of the excavation. The dashed lines indicate what the corresponding elastic stresses and displacements in the absence of a fracture zone would have been.

a function of the distance x ahead of the slit and that f(x) describes the convergence across the fracture zone normal to the plane of the slit. For an increase, $\Delta x = \Delta c$, in the width of the slit the work done by the rock mass on the fracture zone at both ends of the slit is given by:

$$\Delta W = 2 \int_0^\infty F(x) \cdot \Delta x \cdot \{f(x-\Delta x) - f(x)\} + 2SU\Delta x \quad (3)$$

where U = the strain energy in the element of rock Δx that is removed as the slit is enlarged, and

S = the thickness of rock removed to enlarge the slit.

Now

$$\frac{f(x-\Delta x) - f(x)}{-\Delta x} = f'(x), \quad (4)$$

so that,

$$f(x-\Delta x) - f(x) = -f'(x)\Delta x \quad (5)$$

or,

$$\Delta W = -2 \int_0^\infty F(x)f'(x) \, \Delta x \Delta x + 2SU\Delta x, \quad (6)$$

that is,

$$\frac{\Delta W}{\Delta x} = -2 \int_0^\infty F(x) \, f'(x)\Delta x + 2SU = \frac{\Delta W}{\Delta c}, \quad (7)$$

or,

$$\frac{\Delta W}{\Delta c} = 2 \int_{f(\infty)}^{f(o)} F(x) \, df + 2SU. \quad (8)$$

The integral in (8) is merely twice the work done in compressing the rock in the fracture zone at each end of the slit through the displacements to which it is subjected as the slit is enlarged.

Equations (2) and (8) can be used to estimate the value of c* when a fracture zone first forms and the height of the fracture zone, for values of c greater than this critical value, if the appropriate values for the work done in compression and the strain energy are known.

The value of the specific strain energy in each volume of rock removed from the two ends of the slit as it is enlarged, Δc, is the specific strain energy in this rock subjected to a stress equal to its uniaxial compressive strength in plane strain, that is,

$$U = \frac{(1-\nu)}{4G} C_0^2 \quad (9)$$

where U = the specific strain energy, and

C_0 = the uniaxial compressive strength of the rock.

From (2), (8), and (9), the critical value of the half width of the slit when the fracture zone forms is given by:

$$c^* = \frac{S}{2\pi} \frac{C_0^2}{\sigma_v^2} \quad . \quad (10)$$

For many practical situations the value of c given by (10) is so small that the case without a fracture zone is unimportant.

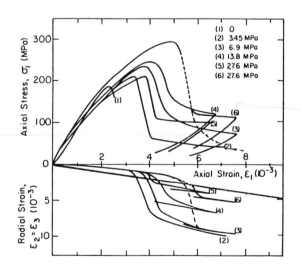

Figure 2. Examples of complete axial stress-strain and axial strain-radial strain curves for specimens of quartzite tested in triaxial compression. The dashed lines indicate the path for deformation of rock in the fracture zone.

3

In Figure 2 are shown typical complete stress strain curves for hard rock in triaxial compression[5]. Once a fracture zone forms the transition from elastic to fractured rock occurs some distance ahead of the end of the slit, close to the peak of the vertical stress illustrated in Fig. 1. The fractured rock between the end of the slit and this peak provides confining stress to the rock undergoing this transition, so that failure occurs in triaxial compression at a vertical stress greater than that corresponding to the uniaxial compressive strength of the rock. The magnitude of the triaxial confinement in the fracture zone decreases from a maximum, at the transition, to zero at the end of the slit. The stress strain path through which the rock in the fracture zone is compressed crosses individual complete stress strain curves for constant confining stresses. It must begin at the maximum strength of the rock in the fracture zone, corresponding to the maximum value of the confining stress in this zone, thereafter passing through points on complete stress strain curves corresponding to ever decreasing confining stress and increasing axial strain. One such path is illustrated by the dashed lines in Figure 2. The specific work done in traversing this path is about one megajoule per m^3, compared with about 0.2 MJ per m^3 in uniaxial compression. Note also the rapid dilatation that occurs in traversing this path as is shown in the $\varepsilon_2 = \varepsilon_3, \varepsilon_1$ portion of the Figure.

This dilatation produces a net increase in specific volume of the rock in the fracture zone, given by $\varepsilon_1 + \varepsilon_2 + \varepsilon_3$, that can be accommodated only through inelastic volumetric closure of the slit, that is, the excavation. The fracture zone is known to extend into the roof and floor, and the larger the fracture zone the larger is the total volume of dilatation that must be accommodated. Let the total height of the fracture zone be H and the specific work done in compressing the rock in the fracture zone, as illustrated in Figure 2, be V, and assume that the strain energy of the fractured rock removed at the edge of the slit is negligible. From equations (2) and (8) it follows that:

$$\pi c (1-\nu) \sigma_v{}^2/G = 2VH, \qquad (11)$$

or

$$\frac{H}{c} = \frac{\pi \sigma_v{}^2(1-\nu)}{2VG} \qquad (12)$$

For a depth of 2km, or σ_v of about 50 MPa, G of about 50 GPa, and V of about one MJ/m^3, H/c is of the order of 0.06.

The stresses in the vicinity of the end of a long thin slit or a very narrow elliptical opening are approximated well[6] by:

$$\sigma_1 = (c/2r)^{\frac{1}{2}} p \cos \tfrac{1}{2}\theta \ (1+\sin\tfrac{1}{2}\theta) + \tfrac{1}{2}p$$
$$\sigma_2 = (c/2r)^{\frac{1}{2}} p \cos \tfrac{1}{2}\theta \ (1-\sin\tfrac{1}{2}\theta) + \tfrac{1}{2} \ , \ (13)$$

where r and θ are local radial coordinates measured from the end of the slit, and the vertical component of the virgin stress $\sigma_v = p$ and the horizontal component $\sigma_h = p/2$. The strength of hard rock can be approximated by a Coulomb criterion of the form

$$\sigma_1 \geq C_0 + q\sigma_3 \ , \qquad (14)$$

where C_0 = the uniaxial compressive strength and q is a pressure hardening coefficient. Substituting equations (13) in (14) yields:

$$\frac{r}{c} = \tfrac{1}{2}\left[\frac{\dfrac{C_0}{p} + \tfrac{1}{2}(q-1)}{\{\cos\phi(1+\sin\phi) - q\cos\phi(1-\sin\phi)\}}\right]^{-2}$$

where $\phi = \tfrac{1}{2}\theta$. The maximum value of r out to which fracture can be extended is determined by the maximum value of $\{\cos\phi(1+\sin\phi) - q\cos\phi(1-\sin\phi)\}$. This expression has a maximum that depends mildly on q. For q = 3 this occurs at $\phi = 58°$ and for q = 5 it occurs at $\phi = 63°$; the corresponding values for the expression are 0.732 and 0.61 respectively. For a depth of 2km or $\sigma_v = 50$ MPa, q = 3 and $C_0 = 200$MPa, r/c is only about 0.01, so that for a slit of width 2c = 500m the height of the fracture zone would be only H = 2r = 5m. This is only a third of that needed to dissipate the released energy through deformation of the fracture zone, given by (12). Nevertheless, the direction $\theta = 2\phi$ that yields the maximum value of $\{\cos\phi(1+\sin\phi) - q\cos\phi(1-\sin\phi)$, or the locus of minimum Coulomb strength, is probably a preferred direction for fracture near the end of a slit-like excavation. Another preferred direction may well be that of the locus of maximum stress difference, that is, $\phi = \pi/4$, as can be seen from equation (13).

FRACTURES

It is now possible to present a plausible hypothesis concerning the origin and development of fractures in the fracture zone ahead of a slit-like excavation.

Two types of fractures can be expected to occur in the vicinity of the ends of the excavation; first, confined shear fractures and second, low confinement cleavage fractures. The dilatancy associated with fracture diminishes rapidly with increasing confining stress, Figure 2, so that most of the deformation which results from confined shear fractures can be accommodated by elastic deformation of the rock around them. Confined shear fractures can, therefore, propagate away from the intersection of the roof (or floor) and the face of the excavation into the solid rock above (or below) the excavation along a direction close to that of the locus of minimum Coulomb strength. Fractures at low confining stress are very dilatant and can occur only adjacent to free faces of the excavation, where the dilatant volume expansion of the rock can be accommodated by deformation of the rock into the excavation. High dilatancy cleavage fractures, therefore, can form only parallel to the free face of the excavation, or any similar free surface formed as a result of fracture.

The distance to which confined shear fractures can propagate into formerly solid rock depends upon the extent of stress concentrations around the excavation of magnitudes sufficient to cause fracture. The extent of these stress concentrations is primarily a function of the width

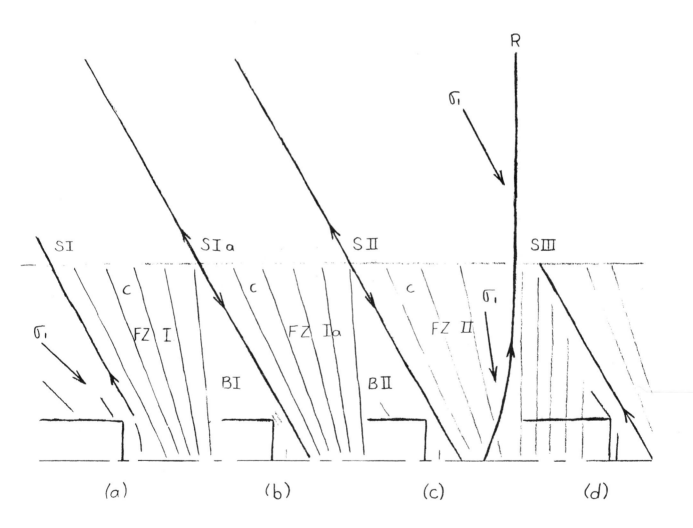

Figure 3. A cross section through the end of a slit-like excavation illustrating the nature of the fracture zone and the three types of fracture that form it, namely, inclined shear (S), cleavage (C), and vertical shear (R). Note the directions of the arrows on the shear fractures which indicate the probable direction of propagation of these fractures; and their orientation with respect to the directions of the maximum principal stress, σ_1.

of the excavation, and the magnitudes of them depend mainly upon the depth of the excavation below surface. As the width of an excavation increases, shear fractures can be expected to propagate to increasing distances from the excavation. The trajectory of the maximum principal stress behind the edge of the excavation is inclined at a greater angle to the vertical than is the locus of minimum Coulomb strength, so that the sense of relative motion across a shear fracture along this locus causes the roof (or floor) behind such a fracture to move into the excavation. The effect of this is to diminish greatly the confining stress in the rock ahead of such a shear fracture. This enables a series of cleavage fractures to form a fracture zone, FZI, ahead of the shear fracture at the face of the excavation, Figure 3a. Successive cleavage fractures are likely to be orientated ever closer to the vertical as they tend to follow the direction of the maximum principal stress in this region. The dilatation of these cleavage fractures restores confining stress to the solid rock ahead of them, so that high stress concentrations develop in this rock. At some stage, these stress concentrations become great enough to produce another shear fracture, SIa. However, it is unlikely that this shear fracture will originate from the intersection of the roof and face of the excavation, where cleavage fractures now exist. Rather, it is more likely to originate in the vicinity of the end of the last cleavage fracture, particularly if this should coincide with a pervasive horizontal weakness such as a bedding plane, Figure 3b. This shear fracture forms a triangular block of solid rock, BI, between it and the last of the series of cleavage fractures. The high stresses that must have existed in this block before the shear fracture developed are partially relieved by the shear deformation, and a new zone of cleavage fractures, FZIa, forms ahead of the shear fracture. As the excavation is advanced into the block, BI, the remaining stresses in it and the confining stresses ahead of it, resulting from the dilatation of the cleavage fractures, are relieved and a new shear fracture SII forms, Figure 3c, producing another block of solid rock, BII and another zone of cleavage fractures, FZII,

ahead of it. This sequence of fracturing may be repeated many times as the excavation is advanced. However, there is another unique type of fracture that warrants consideration.

The locus of maximum stress difference in the rock around an excavation is a line from the end of the excavation and perpendicular to the plane containing the excavation. Although the stress differences are greater along this line than anywhere else, the rock is less likely to fracture along this direction than it is along that of minimum Coulomb strength, because the confining stresses are greater across the locus of maximum stress difference than they are across that of minimum Coulomb strength. However, if the zone in which the series of cleavage fractures occurs is allowed to move into the excavation, thereby diminishing the magnitudes of the confining stresses ahead of the face and across the locus of maximum stress difference, it is possible that another type of shear fracture may propagate from the plane of the excavation out along this locus. Such a shear fracture probably would originate in the zone of highest stress concentration ahead of the face inclined opposite to the cleavage fractures, so that the maximum principal stress parallel to these fractures is inclined to the shear fracture, and tending toward the direction of the locus of maximum stress difference, that is, perpendicular to the plane of the excavation, Figure 3d. As it propagates, because of the large initial stress differences across this fracture the shear dislocation across it is also large. The elastic displacements resulting from this shear dislocation cause the rock above the roof (or below the floor) to move direct into the excavation, and the magnitudes of the sress concentrations in the solid rock ahead of it to decrease significantly.

These unique features of this perpendicular shear fracture make it a likely candidate for a rockburst. The perpendicular shear fracture interrupts the development of a fracture zone comprising successive inclined shear and cleavage fractures. Presumably, cleavage fractures form ahead of the perpendicular shear fracture but their length diminishes rapidly with increasing distance into the rock because of the absence

6

of a free face. As the face is advanced under this zone of cleavage fractures, it seems likely that a new inclined shear fracture will propagate outwards from the intersection of the roof (or floor) and the face into solid rock close to the direction of the locus of minimum Coulomb strength, thereby initiating the development of another fracture zone.

The extent of the stress concentrations associated with an excavation increases in proportion to the width of the excavation as does the rate of energy release (provided that the thickness of the excavated rock is sufficient to prevent contact from occuring between the roof and the floor). By similitude, the extent of the fractures, and of the fracture zone which they produce, should also each increase linearly with the width of the excavation. However, the amount of energy absorbed in the formation of the fracture zone should increase only as the height of the fracture zone, Equation (12), so that the height of the fracture zone should increase linearly with the rate of energy release, but the density of fractures in it should remain constant. Finally, the kinetic energy released by unstable fractures would be expected to be proportional to the square of their length. Therefore, the magnitudes of seismic events produced by unstable fractures should increase as the square of the width of the excavation or the rate of energy released by it.

DISCUSSION

In 1966, Pretorius[6], a very experienced mine manager, wrote:

"To my mind this aspect of the problem (fracture observations and stress measurement) is one of the most important and the key to the ultimate solution of the rockburst problem may well be found in a more comprehensive study of the phenomena of rock fracturing. The fracture pattern, as exposed by stoping, records the effects of energy releases and if we could but interpret the record we would be able to reconstruct the mechanism of the occurrences.

"Fracture planes exhibit certain characteristics which allow of their ready classification into two main categories, viz:

1. Normal fracturing, where no harmful effects on the stope walls are apparent.

2. Abnormal fracturing which can be further classified into two sub-divisions, viz:

    (a) Fracturing associated with bad

hanging wall conditions and falls of hanging wall;

    (b) Fracturing associated with rock-bursting . . .

1. (Normal Fractures) usually dip at 60° to 80° toward the face in the hangingwall and away from the face at a similar angle in the footwall. In either case they may be vertical . . They run parallel to the general line of the face and reproduce medium-sized irregularities in face shape as smooth curves . . . They are normally ½ in. to 6 in. (12 mm to 150 mm) apart (and) . . . vary from hair line cracks to about ¼ in. (7 mm) in width . . . The filling usually consists of finely fractured rock particles and splinters. Powdering is rare and evidence of movement is absent.

2. Two types of (abnormal fractures) are recognized, the first being fractures dipping towards the face in the hangingwall at a low angle of 20° to 40° from the horizontal. This commonly occurs when a stope is open for about 50 to 200 ft (15 m-60 m) at depth . . . The second type of abnormal dip is when fractures dip away from the face on the hangingwall at a steep angle . . . (Fractures associated with rockbursting) are exposed when a face is mined some 4 to 8 ft (1.2 m to 2.4 m) after a rockburst occurred and differ markedly from the two types described. The dip is, more often than not, vertical. The fractures . . . create the impression of being superimposed upon an existing set of fracture planes . . . (The width of fracture) is usually the most significantly different from normal fracture planes. The plane has a width of 2 to 6 in. (50 mm - 150 mm) . . . the filling at burst fractures is highly comminuted or even pulverized. Evidence of movement in the plane of the fracture can commonly be found by 'slickenside' markings on the walls of the fracture. When the burst fracture can be traced in adjacent excavations actual displacement along the fracture can be observed.

"Only a mechanism as described above adequately accounts for the typical burst fracture, which is, in fact, a true shear fracture with movement markings on the planes and comminution or pulverization of the included rock matter. Its near vertical dip and jagged trace are further signs of a fracture developed by a different mechanism to the normal stope fractures."

In terms of the model discussed previously, Pretorius' normal fractures should correspond to cleavage fractures, the first type of abnormal fractures should correspond to SI and SII inclined shear fractures that initiate formation of the fracture zone and the second abnormal type to R vertical shear fractures.

Recently, Adams et al.[7] described three types of fractures:

"Type 1. Steep fractures parallel to the stope face without any displacement in the plane of the fracture . . . are the most common . . .

their dip is essentially vertical . . . The fractures are planar and characteristically cut directly across larger quartz pebbles and grains rather than following the matrix around these grains. . . The dip of the fractures decreases from near vertical in the plane of the stope to about 55°, at a distance of 45 m in the footwall. The density of fracturing also diminishes with distance from the plane of the reef.

"Vertical displacements of up to 50 mm occur across Type 1 fractures. This movement, which is accompanied by the comminution of quartzite along the fracture plane, is however a secondary phenomenon occurring after the formation of the fracture.

"Type 2. Inclined fractures parallel to the stope face with a component of displacement in the plane of the fracture . . . are mining induced normal faults which dip between 60° and 75° both towards and away from the stope face and often occur as conjugate pairs. Displacements of up to 140 mm have been observed across these faults. The fractures revealing the larger displacements consist of both broken and finely comminuted quartzitic material and are several centimeters wide. . The slip direction, as indicated by the smearing out of rock-powder and slickensides, is at right angles to the plane of the reef. This suggests that the maximum principal stress is perpendicular to the plane of the stope and not vertical. . .

"In the footwall, Type 2 fractures which dip in the direction of face advance predominate while in the hangingwall the fractures which dip in the opposite direction are the most common. These fractures extend a short distance through the plane of the stope so that both orientations are equally represented in front of the stope resulting in pairs of conjugate fractures. As the displacement of these fractures is in the same sense as a normal fault, the net result is to allow rock ahead of the stope to displace horizontally into the excavation and also for footwall and hangingwall material to move along these planes into the stope. Thus, this type of fracturing provides a mechanism for rock to move into the mining excavation.

"Type 3. Low angle and vertical, younger fractures form close to the stope face within rock which already contains Type 1 and Type 2 fractures. They are a later generation of fractures, caused by a localized, secondary fracture process. Thus they are superimposed on to the more regional fracture pattern which is developed several meters ahead of the face. The fractures do not occur throughout the stope but develop preferentially in rock which is not initially highly fractured. They have planar surfaces and do not reveal any movement across the fracture plane.

"The fractures are symmetrically disposed around the reef plane, in the hangingwall they dip at 30° and 40° in the direction of face advance and in the footwall dip against the direction of face advance. They do not extend more than 3 m into the hangingwall or footwall and seldom have a strike length greater than 5 m. . .

"It has been found that the distance to which the rock is fractured ahead of the stope face is dependent on the energy release rate. At an energy release rate of 20 MJ/m^2 fracture zones

extend 2 m to 3 m ahead of the face and at 45 MJ/m^2 they extend up to 6 m.

"The density of fracturing in fracture zones however is not increased at higher energy release levels. The average fracture spacing in fractured zones is 43 mm throughout the range of energy release rates analyzed, that is between 5 MJ/m^2 and 75 MJ/m^2."

With some modifications, Type 1 fractures may correspond to cleavage fractures. Cleavage fractures tend to become parallel to the direction of the maximum principal stress so that their dip would decrease with distance from the stope. Type 2 fractures may correspond to vertical shear fractures - in Figure 2 that accompanies their description Adams et al. show that the dips of Type 2 fractures increase with distance from the stope.

Type 3 fractures may correspond to inclined shear fractures SI and SII that propagate away from the stope. Adams et al. claim that they are "later generation" fractures. However, they do not use any evidence that precludes them from occurring at any time, provided that they propagate away from the stope in relatively solid rock. In fact, it is probably most correct to classify them as inclined fractures that initiate in solid blocks of rock subjected to high stress concentrations at the intersection of the face and the roof (or floor).

CONCLUSION

A model of the fracture zone and the mechanism of rockbursts is needed if the origin of rockbursts is to be understood properly and effective steps are to be taken on an informed basis to control or predict them.

The model proposed in this paper should be regarded as a tactic in endeavors to resolve the problem of rockbursts in deep mining. It is the only one of more than a score of different models, based on continuum or discrete behavior of rock in the fracture zone, that withstood comparison against what is currently known about fracture zones and the properties of rock. Nevertheless, it is not likely to prove to be entirely accurate nor complete. However, it is important to have it as a working hypothesis that can be modified, rejected, or supported in the light of additional information. Such a model serves as a needed common framework and focus for the evaluation and utilization of all knowledge about rockbursts. Without such a model it is not practicable to combine information from different sources, such as numerical analysis, physical experiments, and field observations.

REFERENCES

1. Milne, W. G and Berry, M. J. (1976). Induced seismicity in Canada. Engineering Geology, vol. 10, pp. 219-226.

2. Hill, F. G. and Deukhaus, H. G. (1961). Rock mechanics research in South Africa with special reference to rockbursts and strata movement in deep level gold mines. Transactions Seventh Commonwealth Mining and Metallurgical Congress, South Africa, vol. 2.

3. Obert, L. and Duvall, W. I. (1967). Rock Mechanics and the Design of Structures in Rock. New York, Wiley, pp. 582-611.

4. Walsh, J. B. (1977). Energy changes due to mining. International Journal of Rock Mechanics and Mining Sciences, vol. 14, pp. 25-34.

5. Hojem, J. P. M., Cook, N. G. W., and Heins, C (1975). A stiff, two meganewton testing machine for measuring the work-softening behaviour of brittle materials. South African Mechanical Engineer, vol. 25, pp. 250-270.

6. Jaeger, J. C. and Cook, N. G. W. (1979). Fundamentals of Rock Mechanics, 3rd ed., London, Chapman, and Hall, p. 275.

7. Pretorius, P. G. D., contribution to Rock mechanics applied to the study of rockbursts. Journal of the South African Institute of Mining and Metallurgy, vol. 66, pp. 705-713.

8. Adams, G. R., Jaeger, A. J., and Roering, C. (1981). Investigations of rock fracture around deep level gold mine stopes. Proceedings of the 22nd U.S. Symposium on Rock Mechanics, M.I.T., pp. 213-218.

Rockburst hazard and the fight for its alleviation in South African gold mines

M.D.G. Salamon Dipl. Eng. (Min.), Pr. Eng., F.S.A.I.M.M.
Research Organization, Chamber of Mines of South Africa, Johannesburg, South Africa

SYNOPSIS

Rockbursts were first noted on the Witwatersrand around the turn of the century. With the ever increasing scale of mining and the deepening of mines in subsequent decades the danger has become progressively more pressing. Mining engineers used their ingenuity to combat the problem, but around the early 1950's it became clear that the situation could not be alleviated by conventional means and the decision was reached to initiate systematic and large-scale research. This paper is a review of the progress made during the last 30 years and it provides a summary of the present state of the fight against the hazard.

Firstly, an account is given of the nature, frequency, magnitude and severity of the bursts. Next, the mathematical modelling of the behaviour of the rock mass is reviewed. This provides the foundation for the suggestion that a seismic event, which may or may not become a rockburst, results from the disturbance of an unstable state of equilibrium.

There is both theoretical and experimental evidence to indicate that certain energy quantities are good measures of the potential rockburst risk. The control of these energies is the basis of most methods for the alleviation of the hazard.

The practical defence against rockbursts is based on a combined application of three concepts. These are:

(i)   effective face support,
(ii)   good layout design and
(iii)   control of convergence volume.

The paper reports on the progress which has been made in developing practical embodiments of all three of these concepts.

Finally, a brief discussion is given of the current and future research efforts directed towards the amelioration of the rockburst problem.

INTRODUCTION

Rockbursts are perhaps the most feared and persistent threat to the security of miners who operate in certain mining areas. Unfortunately, the goldfields of the Witwatersrand system belong to those mining regions of the world where rockbursts have been a major problem for several decades. In 1979, for example, some 62 per cent of all fatalities could be attributed to rockfalls and rockbursts.

A disconcerting feature of rockbursts is that they defy conventional explanation. Hazards arising from the explosion of gas, coal dust or explosives, fires, water inrushes, the use of mechanical or electrical equipment, and so on, are all within the experience of modern man, but this is not so with rockbursts. Another problem is that a man working underground in a rockburst-prone mine may feel exposed and defenceless. There appears to be nothing he can do to diminish the risk to which he is exposed. It is the duty of mining engineers in collaboration with others and with the backing of the mining companies to alleviate this state of affairs.

In this paper a brief account is given of the efforts made to achieve an abatement of the rockburst problem on the Witwatersrand. The kernel of the results emerging from extended

11

research is an understanding of the mechanism of bursts providing guidance for the development of practical measures to combat the problem. An attempt is made throughout the paper to emphasize the mining point of view. This is done firstly to facilitate the application of measures described and secondly to ease the evaluation whether the progress made on the Witwatersrand is applicable to other mining areas.

HISTORICAL REVIEW
Rockbursts on the Witwatersrand
The rockburst problem emerged less that two decades after the commencement of mining on the Central Rand. As early as 1908 the Ophirton Earth Tremors Committee concluded that the tremors observed were associated with mining and attributed their origin to the 'shattering of support pillars'. With the spread and increasing depth of mining the number of tremors observed by the Union Observatory increased rapidly from seven in 1908 to 233 a decade later.[1] The government felt impelled on three subsequent occasions, in 1915, 1924 and 1964, to appoint further committees to investigate the matter.

Recently a tabulation was compiled[2] of average annual fatalities caused by rockbursts and rockfalls during the past five decades as a percentage of accidental deaths from all sources, Table 1.

The disturbing feature of the figures in Table 1 is that while the total fatality rate decreased appreciably during the past 50 years, the same cannot be said for the rate of accidental deaths due to rockbursts and rockfalls, which has remained virtually unchanged during the last four decades.

It might be argued that this static position represents an improvement because during the intervening years the depth and extent of mining have increased significantly. Clearly, however, there is no room for complacency.

Early efforts to combat rockbursts
The eight decades which have passed since the appearance of rockbursts in South African gold mines can be divided into two periods. During the first period, which finished at about the end of the 1940's, the gold mining industry and the appropriate government authorities in fighting the threat relied mainly on the traditional approach of engineers: observation, experience and reasoning, followed by practical trials.

The first three government-appointed committees functioned during this period. It is fascinating to read their reports, which appear to reveal, especially with hindsight, a surprising measure of apparent understanding of rockbursts. Several of their recommendations seem to have withstood the test of time. For example, the 1908 Committee recognized that the shattering of support pillars might be the cause of many of the tremors and consequently cautioned against the use of solid pillars. (This is not to be confused with 'stabilizing' pillars which are discussed later.) The 1915 Committee suggested that inclined shafts should be overstoped even if the area was 'unpay' and the adjacent areas should be sandfilled. Due to space limitations it is not possible to discuss these reports in depth, but those who have a historical interest will find a reasonably detailed analysis of them in a report published by the Chamber of Mines of South Africa.[2]

Table 1. Fatalities due to rockfalls and rockbursts in relation to total fatality rates.[2]

Decade	Annual average fatality rates		
	All sources (per 1000 employees)	Rockbursts and rockfalls	
		(per 1000 employees)	% of total
1926-35	2,36	0,93	39,5
1936-45	1,68	0,74	44,1
1946-55	1,55	0,71	45,9
1956-65	1,44	0,72	50,1
1966-75	1,31	0,73	55,7

Perhaps the most noteworthy development of this period was the introduction of the longwall method. The 1924 Committee already proposed that this method should be considered.[2] Many years had to pass, however, before the system was introduced as a measure to combat rockbursts at East Rand Proprietary Mines Limited (E.R.P.M.)[4] which later spread to several other mines.

It became more and more apparent by the early 1950's that purely practical attempts to solve the rockburst problem were inadequate. The solutions recommended have always been subject to compromise and even contradiction. For instance, support methods advocated to minimize rockbursts have ranged from wastefilling (backfilling) as solidly as possible to complete caving of the worked-out area.[1] A modern reader of these conflicting proposals might be tempted to applaud the author of one and ridicule that of another. The reality is that the mining engineers of the time had no way of knowing which advise to accept and which to reject. Thus, it became evident that scientific research was necessary to obtain both a qualitative and quantitative understanding of the problem. The realization of the need for formal research might be taken to mark the start of the second period of the long fight against the rockburst threat.

In 1953 the Chamber of Mines assumed sponsorship of all rockburst investigations. Firstly, a team of scientists from the Council for Scientific and Industrial Research was recruited. From the early 1960's the Chamber's own Research Organization started to play an increasingly important role in research and gradually assumed overall responsibility. Thus, almost exactly 30 years ago a major research programme was initiated with the specific purpose of working towards the solution of the rockburst problem in the gold mining industry. Although considerable progress has been made, a complete solution has not been found and the effort shows no sign of abating. It is suggested that the research programme, with its specific goal and which has been sustained already for three decades, is unique in the relatively brief history of rock mechanics.

The modern fight against rockbursts
It is useful to distinguish four phases in the effort that was started 30 years ago. These phases

are not entirely contiguous, but overlap to a lesser or greater extent. They can be defined as follows:

(i)   initial research,
(ii)   development of countermeasures,
(iii) implementation of protective measures, and
(iv)   further research.

A brief account of the activities involved in only the first three phases is given in this section. A more comprehensive discussion of future research in phase (iv) is to be presented towards the end of the paper. As phase (iii) falls outside the scope of the present account implementation of countermeasures is discussed only in this section. However, at least one other paper at this symposium deals with practical applications. The main body of this paper is therefore devoted to recounting the progress made in phases (i) and (ii).

Initial research
Phase (i) was characterized by four approaches. The first of these was to collect data through observation of the occurrence of rockbursts and to subject that to statistical analysis. This work aimed to establish relationships between mining variables and rockburst observations.[1]

Concurrently with observations, an intensive laboratory investigation commenced with the purpose of clarifying the behaviour of rock under stress. The intention was to unfold the deformation characteristics and the mechanism of failure of brittle rock.[1,5,6] The work gained tremendous impetus when Cook[7] and Deist,[8] apparently independently, put forward the notion that brittle rocks, under certain loading conditions, can maintain a 'strain-softening' behaviour.

In the late 1950's Cook embarked on an intensive seismic investigation, using an underground network of seismometers.[1,9,10] Perhaps the most striking feature of the early seismic data was that the number of observed seismic events greatly exceeded that of reported rockbursts. These early observations led to the terminology which refers to 'seismic events' and to 'rockbursts'. The definition is that all rockbursts are seismic events but not all seismic events become rockbursts. Rockbursts belong to that sub-set of seismic events which causes damage to mine workings.

Independently of the work already mentioned, some mathematical investigations were in progress [11,12,13] with a view to establishing whether the theory of elasticity would be able to describe the behaviour of rock surrounding excavations in gold mines. At about the same time, extensive campaigns of underground displacement observations commenced. The comparison of the measured and computed components of displacement revealed that the elastic theory is a valid model for the simulation of the behaviour of most of the rock mass surrounding the excavations. Only the fractured rock immediately around the mining cavity fails to behave as an elastic medium and needs to be excluded from the analysis.[1]

Development of countermeasures
During the second phase of the programme practical procedures were developed for implementing the findings of earlier research. There are essentially two approaches to the solution of the rockburst problem, which might be termed 'strategic' and 'tactical'.

The aim of the strategic approach is to diminish the likelihood of encountering rockburst-prone situations in mining. The common features of the strategic means of combating the rockburst hazard are firstly that they involve decisions which affect the mine in the long term and secondly that the beneficial effects become significant only after the mining of a large area.

The purpose of the tactical approach is to ensure that the destructive effect of a seismic event, if it occurs, is minimized. Essentially, improvements to face and to gully supports fall into this category. It is important to note that the tactical means of 'defence' tend to become effective virtually as soon as the improved supports are installed but, of course, their wide-scale introduction takes time.

It may be argued also that the strategic means of defence is aimed at a reduction in the number and severity of seismic events. At the same time, tactical steps seek to diminish the chances that a seismic event will turn into a rockburst or, if it does, hope to abate its severity.

Implementation of protective measures
The development of the various means of countering the rockburst threat has set the stage for the third phase of the industry's programme for improved strata control. This is the phase which is still in progress and will be pursued vigorously in years to come. It involves the industry-wide introduction of measures which were developed earlier or which are still being developed (for example, a systematic and effective backfilling system).

The problems of introducing the new strata control principles and measures were and are tremendous. The primary task is communication and training, involving the organization of many courses to teach the new concepts to various levels of management. The next step is the evolution of a new breed of specialist engineer, the rock mechanics engineer, who can guide the mine management in the choice of support systems.

Progress has been gratifying. Today all major mining groups have a rock mechanics department and virtually all gold mines have their rock mechanics specialists. Planning is done with the aid of computers to compare and rank the hazards resulting from possible mining alternatives. Several new support systems have been developed and introduced,[18] the efficacy of stabilizing pillars has been proven, and so on.

It can be claimed that the fruits of research did not remain unread on the pages of research reports, but that they are being employed to a considerable extent.

RESULTS OF RESEARCH
In this section an attempt is made to summarize the outcome of some 30 years of research. Of course, a brief account is necessarily selective and presented without proof. A reader wishing to obtain a more substantial history will have to consult the references quoted.

Observations and statistical analyses
The earliest quantitative statistical analyses of rockbursts appear to have been initiated by Hill[4,14] in the 1940's when he examined the influence of factors such as layout, shape of abutment and geological weaknesses such as faults and dykes. At the initiation of the

formal research programme in the early 1950's these examples were followed and an information gathering scheme was organized. It was hoped that subsequent statistical analyses of the data would indicate quantitatively the role of various mining factors.[1] Six mines were included in the scheme.

The analyses of the collected data indicated the significance of a number of variables.

  (i)   Excavation size: the rockburst incidence increased with span until about 180 m was reached. When the span increased further, the incidence was reduced from about 0,4 to about 0,3 rockbursts per 1 000 m^2 of reef mined at a span of approximately 270 m and then remained unaffected by span.

 (ii)   Abutment size: the rockburst incidence increased with a decrease in abutment size until an abutment area of about 150 m^2 was reached; it then decreased until the remnant was stoped out.

(iii)   Depth below surface: a significant positive linear correlation was found between rockburst incidence and the depth below surface, provided the data were so categorized that the influence of other parameters was eliminated.

 (iv)   Dykes: the rockburst incidence for stopes through dykes was considerably higher than for those remote from dykes. This effect was especially marked when these dykes were in small abutments.

  (v)   Faults: the effects of faults were similar to those of dykes.

 (vi)   Stoping width: it was found that in comparable stopes an increase in stoping width was in most cases associated with an increase in rockburst incidence. However, it was difficult to determine whether the relationship was causal; a rockburst could result in a fall of hanging, thereby increasing the stoping width.

Most of the relationships mentioned caused no surprise, since miners suspected their signifi-cance from their own informal observations. They also supported the conclusions arrived at later with the aid of modern methods.

It should be noted that the investigators in this study were handicapped by the disability to observe seismic events directly. They had to rely on reported data on rockbursts, which are generally less reliable and affected by many additional factors.

Seismic investigations

Seismic records of tremors in the Johannesburg area have been kept since 1910. Several investi-gators have examined these data and have noted several interesting features.[15] As early as 1920 Cazalet reported a peak in the diurnal distribution of events on weekdays at about blasting time. There is also a weekly cycle in the incidence with its maximum on Thursdays and its minimum on Sundays. Already in 1946 a direct relationship was established between mining and tremors. In 1971 some 1 600 events were recorded in Pretoria by equipment of the World Wide Standard Seismographic Network. These tremors, which originated on the mines of the Witwaters-rand system, had magnitudes ranging from 2,0 to 4,2 on the Richter scale.

The first three-dimensional underground array of seismometers was employed by Cook in a pioneering investigation into the rockburst problem.[9] This type of network has proved to be the most effective tool in unravelling the complexities of seismicity associated with mining.

Perhaps the most important result of Cook's study was the realization that only a small fraction of seismic events causes damage to mine workings, that is very few seismic events manifest themselves as rockbursts. The relevant part of his data is given in Table 2. Note that the percentage of rockbursts rises with increasing magnitude of the seismic events.

Table 2. Seismic events and rockbursts[1]

Energy of event (ft lb)	10^4	10^5	10^6	10^7	10^8
Number of seismic events	278	121	36	8	2
Instances of damage or number of rockbursts	0	2	2	1	2
Percentage of bursts	0	1,8	5,6	12,5	100

The investigations of Cook and others showed that most of the foci of seismic events were confined to the immediate vicinity of mining activities, in other words to regions of high stress and of active mining,[15] Figs. 1 and 2.

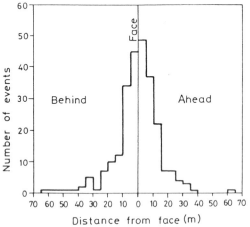

Fig. 1 Histogram showing the position of the face relative to the seismic event, after Cook.[15]

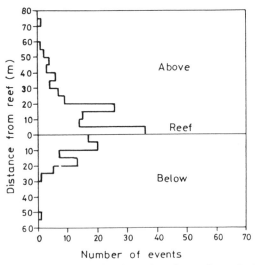

Fig. 2 Histogram showing the elevation of the seismic event relative to the reef plane, after Cook.[15]

A detailed examination of diurnal, Fig. 3, and weekly, Fig. 4, distributions provides further evidence of the close association between seismic events and mining.[9,15] Also, there are clear indications of the time-dependency of the deformation of the fractured rock around excavations.[15]

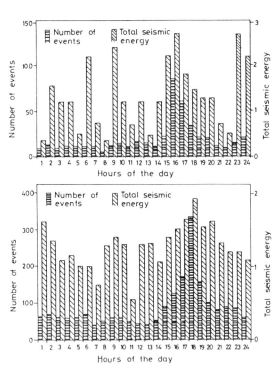

Fig. 3 The diurnal distributions of the number and kinetic energy of seismic events on two mines, after Cook.[15]

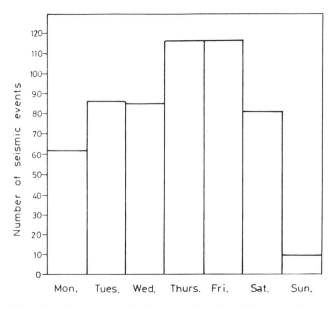

Fig. 4 Histogram of the weekly distribution of seismic events, after Cook.[15]

In more recent years McGarr[16] and others have started to make increasing use of the science of seismology. They present a convincing argument in proposing that there is no fundamental difference between natural earthquakes and mining-related seismic events. Taking advantage of this conclusion, McGarr has proposed a number of practical measures concerning the level of hazard, the maximum expected magnitude of events and the frequency of events producing ground velocity greater than a given value.[16] Although there is uncertainty about the validity of some of the basic premises employed, the approach obviously warrants further study.

Some ten years ago it first became doubtful whether the ambient seismicity in the Klerksdorp goldfield is entirely due to mining induced stress concentrations. Intensive seismic investigation in the area has tended to substantiate the suspicion that this is not the case, suggesting that tectonic forces play a major role and in many instances mining may only act as a trigger mechanism.[17]

It is relevant to mention that two decades of wide-ranging seismic investigations (nine mines operate seismic networks at present) have given independent proof that most of the rock mass reacts as an elastic medium. The propagation of seismic waves provides useful confirmation of this conclusion.

Modelling the behaviour of the rock mass: ✓
elastic medium
The concept of tabular excavation

The gold-bearing conglomerates of the Witwatersrand system show an economic persistence and stratigraphic continuity which are unique among metal-mining regions of the world. The economic horizons or reefs consist of a few centimetres to a few metres of quartzitic conglomerate surrounded by hard quartzites. Thus, the reefs are typical examples of what is known as tabular ore bodies, of which coal seams are perhaps the best known examples.

The common features of tabular deposits are that: (i) they extend over large areas, (ii) they frequently lie virtually in a plane or alternatively in either parallel or non-parallel places resulting from faulting, and (iii) their width is fairly uniform and it is negligible in comparison with the other two dimensions. Naturally, the mining cavities created in the course of mining such deposits, apart from auxiliary excavations, are also tabular.

Considerable progress has been made in the analysis of problems involving tabular excavations since the early work of Hackett in 1959.[19] It would appear that he was the first to suggest that significant simplifications can be achieved when solving problems involving tabular excavations firstly by employing an elastic model, and secondly by neglecting the influence of surfaces which are normal to the plane of the deposit in comparison with the effects of those which are parallel with this plane.

Mathematically this simplification is equivalent to viewing tabular excavations as displacement discontinuities or 'slits', the outlines of which coincide with the plan of the excavations. This means that the coordinates of two points, of which one is on the hangingwall and the other perpendicularly opposite on the footwall, are the same but their displacements are different. The components of the traction vector, however, are continuous when moving from the hangingwall to the footwall.

Naturally, the employment of displacement discontinuities or slits to model tabular excavations introduces errors in the vicinity of the edges of the cavities, but these will not appreciably affect the displacement and stress components elsewhere.

Analyses involving tabular excavations

It was realized as early as 1961[20] that the use of the slit model would yield significant benefits only if it permitted the analysis of practical mining layouts, which are complex and irregular. The expression 'face element' was coined in 1962[21] to describe the basic element of the procedure, which is clearly a forerunner to the modern 'boundary element' method of computing stresses and displacements around mining excavations in an elastic medium. A fairly comprehensive review of methods relevant to the treatment of tabular excavations is given in the General Report on Underground Openings presented at the Third Congress of the International Society for Rock Mechanics at Denver in 1974.[22] Much of what is said in this section is a summary of the relevant parts of that report.

In specifying problems associated with tabular excavations, the relative displacement vector of a point on the hangingwall and that on the footwall opposite plays an important role. Assume that the x_3 axis of the coordinate system is normal to the deposit and points towards the footwall. Let the components of the relative displacement vector be s_i and define them as follows:

$$s_i = u_i^+ - u_i^-, \qquad i = 1,2,3, \qquad (1)$$

where u_i^+ and u_i^- refer to the induced displacements of the footwall and hangingwall respectively. These definitions arise from the convention that s_3 should be _positive_ when it denotes _convergence_ and that u_3 is positive when it represents motion towards the hangingwall. The latter of these statements is a consequence of the basic convention that compressive normal stresses are taken to be positive. It should be noted that components s_1 and s_2 are the _ride_ components in the x_1 and x_2 directions respectively.

Let the ride and convergence components at point ξ_1, ξ_2 in the reef be $s_k(\xi_1, \xi_2)$ with $k = 1,2,3,$. Here ξ_j, $j = 1,2$ represent an auxiliary coordinate system in the reef. The elementary induced displacement components at an arbitrary point x_1, x_2, x_3 in the rock mass can be expressed as follows:

$$\Delta u_i^{(i)} = \sum_{k=1}^{3} s_k(\xi_1, \xi_2) U_i^k .$$

$$\{(x_1 - \xi_2), (x_2 - \xi_2), x_3\} \, \Delta A,$$

$$i = 1,2,3 \qquad (2)$$

where ΔA is a small area centred at ξ_1, ξ_2 over which relative displacements s_k are supposed to exist. The area ΔA was earlier referred to as the 'face element'.

If the total mined-out area in the reef is A, then it is possible with the aid of (2) to calculate the total induced displacements at x_1, x_2, x_3:

$$u_i^{(i)} = \int_A \sum_{k=1}^{3} s_k(\xi_1, \xi_2) U_i^k \{(x_1 - \xi_1), (x_2 - \xi_2), x_3\} dA, \qquad (3)$$

where, of course, $dA = d\xi_1 \, d\xi_2$. Now employ the usual relationship between displacements and strains, e_{ij}, to obtain:

$$e_{ij}^{(i)} = \tfrac{1}{2} \int_A \sum_{k=1}^{3} s_k(\xi_1, \xi_2) \left(\frac{\partial U_i^k}{\partial X_j} + \frac{\partial U_j^k}{\partial X_i} \right) dA, \qquad (4)$$

where the argument of U_i^k is the same as in (3).

If the result in (3) is substituted into the particular Hooke's law appropriate for the rock mass, the stress components at x_1, x_2, x_3 follow:

$$\tau_{ij}^{(i)} = \int_A \sum_{k=1}^{3} s_k(\xi_1, \xi_2) K_{ij}^k \{ (x_1 - \xi_1), (x_2 - \xi_2), x_3 \} \, dA \qquad (5)$$

Clearly, function K_{ij}^k can be expressed as a linear combination of the derivatives of U_i^k using (4), but the relationship will depend on the applicable form of Hooke's law.

The expressions in (3) to (5) subdivide the problem involving tabular excavations into the solution of two separate sub-problems. Firstly U_i^k are derived, usually analytically. These functions are independent of the mining layout so they can be obtained once and used repeatedly later. Functions U_i^k were obtained for many practical situations during the 1960's and 1970's.

The second fundamental problem involves the determination of the distribution of the relative displacement components. This is a formidable task which must be performed on each occasion when a new mining layout is analysed. The first solutions to this problem were obtained in 1964 with the aid of an electric analogue.[23] Subsequently various digital solution methods were evolved [24,25] and this development is still in progress. The formal part of the digital solution is straightforward.

As before, let the total mined-out area in the reef be denoted by A. Also, assume that the region within A where the hangingwall and footwall are supported or come into contact is B and the unsupported part of the excavations is C. Naturally, A = B + C. An example of the boundary conditions in the plane of the deposit defining a set of problems involving tabular excavations supported by backfill may be expressed as follows:

$$\tau_{i3}^{(i)} = (T_i^{(p)} - f_i(s_i)) \quad \text{in B,}$$

$$\tau_{i3}^{(i)} = -T_i^{(p)} \quad\quad\quad \text{in C,} \tag{6}$$

where $f_i(s_i)$ is a function describing the deformation characteristics of the fill.

As an illustration, the solution for the simple instance where the excavations are unsupported, that is B=0, is presented next:

$$-T_i^{(p)} = \lim_{x_3 \to 0} \int_A \sum_{k=0}^{3} s_k(\xi_1,\xi_2) K_{i3}^k \{(x_2-\xi_1),$$

$$(x_2-\xi_2), x_3\} \, dA, \tag{7}$$

where $T_i^{(p)}$ is the primitive traction vector on the plane of the footwall.

The relationship in (7) represents a system of three simultaneous integral equations for s_k. The exact solution of this system has been found in only a few almost trivial cases. In practice (7) is solved numerically by employing summation in place of integration.

From the mid-1960's onwards analogue-cum-numerical and later purely numerical methods were available to compute the displacements and stresses induced by mining activities in a single reef. This capability permitted the pursuance of work devoted to validation of the elastic model. Extensive comparisons of measured and computed displacements have provided reasonably convincing proof of the postulate that the major part of the rock mass behaves as a homogeneous isotropic elastic solid.[1]

Applications of the elastic model
The establishment of the validity of the results summarized in the previous section by the late 1960's opened the door for practical exploitation of the new situation. These were exciting times since this was the first occasion in the history of mining that extensive evidence existed that the behaviour of the rock mass in a particular geological region is predictable quantitatively.

Applications were developed in the following three directions:

(i) some basic theorems of elasticity provided a firm foundation for the formulation of energy theorems concerning the mining process,

(ii) various computer programs, applying the face element principle, facilitated the computation of the effects of stoping on the stopes themselves, on service tunnels, shafts, etc., permitting the rational design of mining layouts,[26] and

(iii) the results in (3) to (5) provide the ideal means of estimating the upper bound of stresses or strains with a view to developing methods for the design of supporting[27] and shaft[28] pillars.

In view of space limitations only the first of these applications is explored further in this paper.

Energy considerations

General results
Mining is a progressive activity. Excavations in mines are changed in shape and made to grow in size with time. It is logical, therefore, to examine the energy changes which come about due to a specific change in the mining geometry rather than to analyse, as others have done, the overall energy transformations which are associated with the total change in layout as mining proceeds from the virgin state to the current geometry.[22,29] The latter approach obscures important fundamental aspects of energy conversion.[22,30]

Two arbitrary states of mining are defined, with the aim of clarifying energy transitions resulting from the additional mining which transfers the geometry from state I to state II. As a consequence of increasing the size and/or the number of mining excavations, displacements are induced in the surrounding rock. Acting through displacements the external and body forces do some work, W. This work is often referred to as 'gravitational' or 'potential' energy change. Also, a certain amount of strain energy is stored in the rock removed during the process of mining, U_m. The sum ($W + U_m$) can be regarded as the energy quantity which must be expended in some manner as a result of the change in mining geometry.

A part of this energy is accounted for by an increase in the energy content of the rock mass surrounding the mine cavities in state II. This increase in stored energy is denoted by U_c.

19

If some or all of the excavations are supported by means which may or may not include backfill, then a certain amount of energy is dissipated in deforming the support, W_s. It is assumed here that the rock mass is an elastic continuum, therefore no energy is consumed in fracturing or in non-elastic deformation. Thus, the energy accountably expended during the transition from state I to state II is $(U_c + W_s)$.

The amount of energy consumed in deforming rock and support cannot be greater than the total energies available. Moreover, because the energy stored in the rock which is extracted during the change in geometry, that is U_m, is not available to strain rock or support, the following inequality must apply:

$$W \geq U_c + W_s \qquad (8)$$

and, because $U_m > 0$ it follows that:

$$(W + U_m) > (U_c + W_s). \qquad (9)$$

It has become customary to refer to the excess energy which needs to be expended in a manner which has not yet been defined as 'released energy' or W_r. On the basis of (8) and (9) it follows that

$$W_r = (W + U_m) - (U_c + W_s) > 0 \qquad (10)$$

and that

$$W_r \geq U_m > 0. \qquad (11)$$

Thus, additional mining is <u>always</u> associated with some release of energy which has to be expended in some manner.

If the new configuration of excavations were to be reached suddenly by removing instantaneously all rock which is to be mined, there would be oscillations in the rock mass. The equilibrium state corresponding to the new mining geometry would be attained through damping, which arises from minor imperfections in the rock. Some energy, the kinetic energy, would be dissipated in the damping process. Denote by W_k the dissipated kinetic energy. Since there is no other mode of dissipation it follows from (11) that

$$W_r = U_m + W_k \qquad (12)$$

and from (10) that

$$W_k = W - (U_c + W_s). \qquad (13)$$

These results, together with the formulae for the various energy components, are discussed in detail in a paper to be submitted for publication.[29]

Although the total energy released during the transition from state I to state II is independent of the mode of transfer, the total values of U_m and W_k are strongly influenced by the number of steps taken to effect the change in geometry. Postulate that the additional mining is achieved in n steps and introduce the following notations:

$$W_k^n = \sum_{i=1}^{n} W_{ki}, \qquad U_m^n = \sum_{i=1}^{n} U_{mi}, \qquad (14)$$

$$W_r^n = \sum_{i=1}^{n} W_{ri} = W_r = \text{constant}.$$

These notations permit the mathematical formulation of the earlier statement, that is:

$$W_r = U_m^n + W_k^n = \text{constant}, \qquad (15)$$

where U_m^n and W_k^n themselves need not be constant. To illustrate this phenomenon the variations in U_m^n and W_k^n for a circular tunnel and for a spherical cavity are plotted in Fig. 5 as a function of the number of steps used to increase the radii of these excavations from zero to the required final value. The theoretical results applied in the calculations were published some years ago.[22]

The following conclusions emerge from Fig. 5:
(i) If the total kinetic energy potential, W_k^n, of creating an underground opening is estimated by assuming that the excavation is made in one step when it is in fact made in $n > 1$, the estimate is grossly misleading. The error increases with the number of steps used to excavate the cavity. For instance, 50,0 per cent of the released energy can become kinetic energy if a spherical cavity is made in one step, but if mining were done in 64 equal steps the kinetic energy potential would be only 3,4 per cent of W_r. There is a corresponding increase in U_m^n.
(ii) More specifically, the curves in Fig. 5 suggest that the magnitude of W_k^n approaches zero as the increment in the radius of the cavity becomes infinitesimal. In fact, as was noted in 1974,[22] when $(r_{i+1} - r_i) \rightarrow 0$ it is correct to say that the energy $W_k^n \rightarrow 0$.

20

The important results arising from the limiting case can be put in the following forms:[22,30]

$$\Delta W = \Delta U_c + \Delta W_s, \quad \Delta W_r = \Delta U_m, \quad \Delta W_k = 0, \quad (16)$$

where the energy quantities are prefaced by the symbol Δ to emphasize that the effects of very small or infinitesimal changes in geometry are considered. It is assumed in deriving (16) that the enlargement is achieved by mining a stress-free portion of the surface of the excavation.

Some fundamental conclusions can be deduced from the results in (16). These are that:

(i) the work done by the external and body forces during a <u>small</u> change in geometry, ΔW, is fully expended in straining the rock mass, ΔU_c, and deforming the support, ΔW_s;

(ii) the energy released in the course of further mining equals the strain energy stored in the rock prior to its removal; and

(iii) the enlargement of mining cavities in small steps is a quasi-static, <u>stable process</u> which does not result in the release of kinetic energy into the rock mass, therefore it cannot be the source of seismic energy.

In the course of mining, cavities are usually excavated in small steps, thus the conclusions listed here are of considerable importance. Of course, the results are exact only in a perfect elastic medium. Such a description does not fit reality where rocks are imperfect and often fractured. Nevertheless, these results represent part of a foundation on which an understanding of the mechanism of rock-bursts can be based.

Tabular excavations

Interest in this paper is concentrated on tabular excavations. In view of this it will be useful to quote in full the formulae for some of the energy components for such cavities. These energy quantities refer to changes arising from the additional mining which transfers state I into state II and were derived elsewhere.[29]

Denote the mined-out area in state I by A_o as well as that portion of A_o which is supported by B_o. In state II the total and supported areas become $A = A_o + A_m$ and $B = B_o + B_m$ respectively. The work done by the external and body forces during the transition from state I to state II is

$$W = (U - U') + W_s + \tfrac{1}{2}\{\int_{A_m} T_i^{(p)} s_i^{(i)} dA$$

$$+ \int_B (1-\alpha_i) R_i^{(i)} s_i^{(i)} dA\}, \quad (17)$$

where W_s is the work done on the support or backfill:

$$W_s = \int_{B_o} R_i^{(p)} s_i^{(i)} dA + \tfrac{1}{2}\int_B \alpha_i R_i^{(i)} s_i^{(i)} dA \quad (18)$$

and

$$U - U' = U_c - U_m. \quad (19)$$

Here U and U' are the total strain energies stored in the rock mass in state II and state I, respectively. With the aid of (19) the released energy in (10) can be suitably rephrased for tabular cavities:

$$W_r = W - \{(U - U') + W_s\}, \quad (20)$$

from which, as well as from (17), the following result follows immediately:

$$W_r = \tfrac{1}{2}\{\int_{A_m} T_i^{(P)} s_i^{(i)} dA + \int_B (1-\alpha_i) R_i^{(i)} s_i^{(i)} dA\}. \quad (21)$$

In these expressions $T_i^{(p)}$ is the traction vector in state I acting on the surface A_m which is to be exposed, while $R_i^{(p)}$ and $R_i^{(i)}$ denote the support resistance vector in state I and the increase in this resistance due to the additional mining respectively. Parameter α_i is related to the shape of the deformation curve of the support as shown in Fig. 6. The value of α_i is chosen in such a manner that $\tfrac{1}{2}\alpha_i R_i^{(i)} s_i^{(i)}$ equals the cross-hatched area in Fig. 6.

Mining of narrow tabular deposits is carried out almost exclusively in small steps. Thus, it is important to transpose the results in (16) to a form appropriate for handling tabular excavations.

In deriving the relationships in (16) no restriction was placed on the shape of the mine opening, therefore they must be valid for tabular cavities also. If the second expression in (16) is applied to (21), it leads to

$$\Delta W_r = \Delta U_m = \tfrac{1}{2}\int_{\Delta A} T_i^{(p)} \Delta s_i dA, \quad (22)$$

where $A_m = \Delta A$ and $s_i^{(i)} = \Delta s_i$ indicating that the increment in the area of mining is small. The second term in (21) is zero because $\alpha_i = 1$ for a small increment. The same substitutions into (17) lead to

$$\Delta W = \Delta(U-U') + \Delta W_s + \Delta U_m = \Delta U_c + \Delta W_s \qquad (23)$$

confirming the first result in (16). It may be noted that the relationship in (22), as will be seen later, has proved to be the basis of one of the most important practical means of fighting the rockburst hazard.

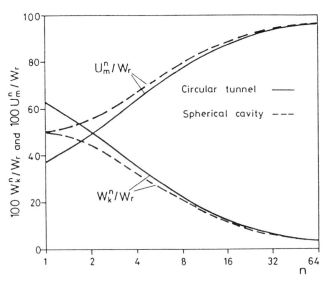

Fig. 5 The split of the total released energy, W_r into total (potential) kinetic energy, W_k^n, and total strain energy (removed with the excavated rock).

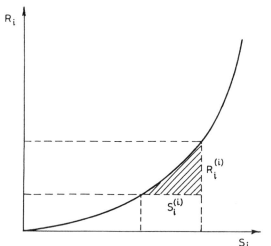

Fig. 6 The work done during the additional compression of backfill and the definition of the parameter α.

MECHANISM OF ROCKBURST

Seismic events
Seismic energy and causes of seismic events
Mining gives rise to seismic activity ranging from micro-seismic events radiating as little as 10^{-5}J to tremors radiating as much as 10^9J.[15] However, the attention is focused in this paper on events which might be the cause of considerable damage to workings, therefore it will be convenient to exclude small incidents from consideration. Thus, a sudden, violent occurrence, radiating a significant amount of kinetic energy, say not less than 10^4J, will be regarded as a seismic event for the purpose of the present discussion.

All seismic events radiate kinetic energy. The identification of the source of this energy is an essential prerequisite to the understanding of the mechanism involved.

It has been widely accepted for some years that the source of kinetic energy is the energy which is unavoidably released in the course of mining.[1,9] Even as recently as 1982, McGarr[16] calculated a 'seismic efficiency' by comparing the seismic energy measured in a given region and time span to the energy released by the corresponding mining activity. Computations based on this concept yield efficiency values in the region of 0,1 per cent.[15] Such low values raise doubt with regard to the validity of the model used.

In fact, the justification for assuming a direct relationship between seismic and released energies disappeared almost a decade ago. It was reported in 1974[22] that theoretically during the enlargement of cavities in small steps, which is the normal course of mining in most cases, the released energy is removed with the excavated rock. Thus, it was concluded that in most mining operations, the released energy cannot be the direct cause of seismic activity.

The question therefore remains: what is the source of the seismic energy?

In seeking an answer, it is necessary firstly to explain that on some occasions a small step in mining can suddenly liberate a considerable quantity of energy. With this purpose in view the nature of the various states of equilibrium which may exist are examined next.

For a system to be in equilibrium its potential energy must have a stationary value. The state of equilibrium is 'stable' or 'unstable' depending on whether the stationary value is a minimum or a maximum. That is to say, while additional external work is required to change a system in a state of stable equilibrium energy is extracted from the system by disturbing a state of unstable equilibrium. Fig. 7 is an illustration of a pair of simple examples of these two types of equilibria.

It can be concluded that the kinetic energy radiated by a seismic event is the manifestation of the energy 'extracted' from the system in the course of upsetting a state of unstable equilibrium. Thus, for a seismic event to occur the stresses in the rock mass must be on the brink of unstable equilibrium.[22]

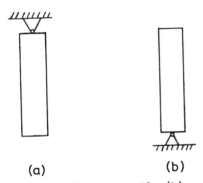

(a)   (b)

Fig. 7 Stable (a) and unstable (b) equilibria

Reasons for instability

Drucker postulated the existence of a family of 'stable continua' in which instability cannot occur.[31] According to his theory it is sufficient for a continuum to be stable if

$$\delta\tau_{ij}\delta e_{ij} > 0, \tag{24}$$

where $\delta\tau_{ij}$ and δe_{ij} are any infinitesimal stress changes and the corresponding infinitesimal changes in strain respectively. Thus, if a continuous material satisfies the inequality in (24) the equilibrium state of any problem involving such a medium will be stable, provided the strains remain 'small'.

Drucker's definition of stable continua includes time-independent and time-dependent, and linear or non-linear elastic or plastic media. To extend the definition in (24) to time-dependent materials it is necessary to replace δe_{ij} by $\delta\dot{e}_{ij}$, where the dot signifies that the time derivatives of strain changes are required. The common features of Drucker's stable media is the property of 'strain hardening'.

Thus, to provide an explanation of the often observed phenomenon of seismic events one or other of Drucker's basic postulate must be relaxed. Either the rock mass has to be (or must become) discontinuous or the rock comprising the mass must become an unstable material when subjected to certain loading conditions. The acceptance of the presence of discontinuities in the mass and the supposition that the rock can become an unstable material represent two separate but complementary lines of attack on the problem of rockbursts. Both of these approaches will be discussed briefly.

Discontinuities. The rock mass might be jointed, faulted or stratified and subjected to gravitational, tectonic and mining induced stresses. Geological features have no influence on the stability of the system as long as their opposing surfaces are bound securely by cohesion and/or friction.

As a result of mining the state of stress at some point in the mass may reach critical values so as to overcome the friction and/or cohesion which bind together the surfaces of pre-existing geological discontinuities. The state of equilibrium immediately prior to this event is obviously unstable.

A simple mechanical model is depicted in Fig. 8 to demonstrate the principle of this type of instability. Consider a block which is placed on a rough surface and which is acted on by a normal force N and a tangential force T. The latter is assumed to exert its influence through an elastic spring. This system will be in stable equilibrium as long as

$$T - \mu_s N > 0, \tag{25}$$

where μ_s is the static coefficient of friction. Thus, once N acts on the block, force T can be increased gradually without moving the block as long as the inequality in (25) holds. When $T = \mu_s N$ the system is in an unstable equilibrium and the block will come into motion due to either an infinitesimal increase in T or an infinitesimal decrease in N. When the block starts to move, the smaller, dynamic coefficient of friction,

23

μ_d, becomes operative and as a result a force of initial value $(\mu_s - \mu_d)N$ accelerates the block which gains kinetic energy in the process.

New discontinuities can be created in the mass surrounding excavations as a result of rock failure.

An example of such an instance is given in Fig. 9 which depicts the slow tensioning of a cylindrical bar.[22] Assume that when the stress reaches the value of the tensile strength, say $-\sigma$, the bar snaps at some cross-section mn. Immediately prior to and after the failure the stress acting on cross-section mn is $-\sigma$ and zero respectively. This change in stress can be achieved by applying suddenly a uniformly distributed compressive stress with a magnitude σ to the freshly exposed surfaces of section mn. The suddenly applied stress, like a hammer blow, initiates the propagation of waves of compression in both parts of the bar. Thus, a 'seismic event' is created.

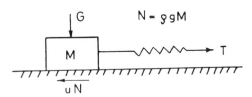

Fig. 8 Mechanical model for a slip on a geological weakness.

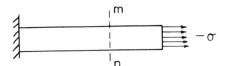

Fig. 9 The creation of a seismic event as a result of the failure in tension of a cylindrical bar of rock

Unstable material. It was recognized during the mid-1960's that many rocks can exhibit 'strain softening' behaviour under certain circumstances with the concomitant property that they are 'stable' only in the sense of (24) in the ascending branch of their stress-strain relationship. Rocks in the strain softening state are only conditionally stable. This means that the criteria for their stability can be formulated only in terms of their own deformation behaviour

and that of the surrounding rock.

Although the behaviour of rocks in the strain softening mode is poorly understood, some practical solutions have been derived. Perhaps one of the most notable examples relates to stability of pillar workings.[24,32,33] General criteria for instability around excavations have not been obtained as yet.

Sources of seismic energy

Having established that the immediate cause of a seismic event is instability, it still remains to consider the problem of the identification of the source of the kinetic energy imparted to the rock particles during an event. It has been suggested previously that this energy must be drawn from the strain energy stored in the rock surrounding the focal region of the event.[9,22,34]

The failure of the bar in Fig. 9 is a good example of the transformation of strain energy into kinetic energy.[22] As a result of the failure, waves of compression begin to travel along both parts of the broken bar with a velocity c. Since the compression waves neutralize the existing tensile stress, after the elapse of time t portions of the bar with a total length of 2ct become de-stressed. The remaining parts continue to be subjected to the tensile stress $-\sigma$.

Both the stored energy recovered from and the kinetic energy imparted to the de-stressed length is $Act\sigma^2/E$, where A is the cross-sectional area. Thus, the waves acquire their energy from that stored in the rock.

Another example of the same principle is provided by the mechanical model depicted in Fig. 8. Here a block starts to slide when the static friction restraining it has been overcome. Immediately prior to the commencement of the motion the strain energy stored in the spring is

$$U_{max} = \mu_s^2 N^2/2k, \qquad (26)$$

where k is the spring constant. Introduce parameters $\psi = \mu_s/\mu_d$ and $\alpha^2 = k/M$. The energy stored in the spring, after the elapse of an interval t since the start of the motion, reduces to:

$$U = U_{max}\{\psi + (1-\psi)\cos\alpha t\}^2 \qquad (27)$$

Thus, an energy quantity, $(U_{max}-U)$, has been liberated as a result of the relaxation of the spring caused by the movement of the block.

The accelerating block, of course, gains kinetic energy which can be expressed as follows:

$$W_k = U_{max}(1-\psi)^2 \sin^2 \alpha t. \qquad (28)$$

Part of the liberated energy is transformed into heat as a result of friction. This heat energy is given by:

$$W_h = 2U_{max}\psi(1-\psi)(1-\cos\alpha t) \qquad (29)$$

It is a simple matter to show that the source of both the kinetic energy and the heat energy is the liberated strain energy, that is:

$$U_{max} - U = W_k + W_h \qquad (30)$$

for all values of t.

These examples demonstrate that usually the source of seismic energy is the energy stored in the rock. In most instances this stored energy is due to body forces and to induced stresses associated with mining. On occasions, however, tectonic forces play an important role and their contribution to strain energy is dominant. In the former case the foci of events are likely to be fairly close to some mining activity, but in the latter this need not be so. This difference has important practical implications which will be discussed later.

It should be noted that strain energy need not be the sole source of seismic energy. For example, the unstable collapse[24,32] of pillars in shallow depths may result in a relatively large loss of potential energy of the rock mass. Significant seismic activity is associated with such a failure[35,36] but this type of event is outside the scope of this paper.

Initiation of seismic events

A seismic event will occur only if a set of conditions pre-exists. It is possible at this stage to formulate the necessary conditions for this to happen.
(i) A region in the rock mass must be on the brink of unstable equilibrium either because
    (a) the presence of an appropriately loaded pre-existing geological weakness such as a joint, fault, dyke or bedding plane; or because
    (b) the changing stresses are driving a volume of rock towards sudden failure; or because

    (c) some support system, for example a system of pillars, approaches a state in which its unstable collapse is imminent.
(ii) Some induced stresses must affect the region in question and the magnitude of these stress changes, however small, must be sufficiently large to trigger off the instability.
(iii) Sudden stress change of sizeable amplitude must take place at the locus of instability to initiate the propagation of seismic waves.
(iv) Substantial amount of energy must be stored in the rock around the instability to provide the source of kinetic or seismic energy. The origin of this stored strain energy is work done by:
    (a) gravitational forces and/or
    (b) tectonic forces and/or
    (c) stresses induced by mining.

Seismic investigations have proved that most of the foci of seismic events are confined to the vicinity of current mining activities, especially if these take place in highly stressed zones. It is necessary to resort to a fundamental feature of seismic activity around mining excavations to elucidate this observation.

No mining geometry or layout appears to exist which unfailingly leads to a seismic occurrence. At best, the likelihood of a seismic event occurring is greater in one situation than in another. Thus, seismicity is a statistically predictable rather than deterministically foreseeable phenomenon. The proneness of a mining geometry to cause a seismic event is a variable which consists of a deterministic and a random component. The presence of the random component requires an explanation.

The occurrence of unstable equilibria in rock is usually associated with the presence of pre-existing 'cracks' (unstable collapse of supports is excluded here). The origin, size, attitude and distribution of these cracks are varied. This variability constitutes the statistical element in seismicity. For the sake of simplicity, it is postulated here that the size, attitude and distribution of cracks are random variables. This means that the number per unit volume of those cracks which cause instability due to a given tensor of triggering stress change, is also a random variable and its value can be expected to grow as the 'amplitude' of the triggering disturbance is increased.

This reasoning leads to important conclusions concerning the frequency of seismic events. The expected frequency rises if

(i) the induced disturbance affecting a given volume of rock increases, and if

(ii) the volume of rock exposed to a given level of disturbance becomes larger.

The stress disturbance induced by a step in mining is primarily determined by (proportional to) the magnitude of stress acting on the surfaces to be exposed and it diminishes with the increasing distance from the region of fresh mining. Thus, it follows that the probability of encountering seismic events becomes greater if a highly stressed part of the reef is mined and the foci of these events are likely to be in the vicinity of the current mining activity.

Since the latter of these statements might not be obvious, the stress changes due to a small enlargement of the span of a horizontal parallel-sided open slit (stope) is analysed next. Let the original span be $2L$ and assume that the face on the right is advanced a small step ΔL. If the virgin vertical stress on the reef horizon is denoted by q then the induced stresses due to the enlargement of the span are as follows:

$$\tau_{11}^{(i)} + \tau_{33}^{(i)} = -q\{\phi'(z) + \phi'(z)\},$$

$$z = x_1 + ix_3,$$

$$\tau_{11}^{(i)} - \tau_{33}^{(i)} + 2i\tau_{13}^{(i)} = q(z-\bar{z})\phi''(z). \quad (30)$$

Here the origin of the co-ordinate system is at the face which is being advanced and axis x_3 points towards the footwall. The stress components are expressed in terms of the complex potential

$$\phi(z) = -\frac{\Delta L}{2}\{1 - (\frac{z+2L}{z})^{\frac{1}{2}}\} \quad (31)$$

For example, the horizontal and vertical principal induced stresses in the plane of the reef are obtained from (30) and (31) and assume the following form:

$$x_3 = 0$$

$$\tau_{11}^{(i)} = \tau_{33}^{(i)} = -qRe\phi'(x_1) = \frac{qL\Delta L}{2|x_1|\{x_1(x_1+2L)\}^{\frac{1}{2}}}, \quad \begin{matrix} x_1 > 0 \\ \\ x_1 < -2L \end{matrix}$$

$$\tau_{11}^{(i)} = \tau_{33}^{(i)} = 0, \quad -2L < x_1 < 0. \quad (32)$$

In Table 3 the stresses calculated from (33) are given for two symmetric positions. The value of the stresses are computed at a distance Δx ahead of the moving face (right) and at a point located in the same position relative to the stationary face (left). In the same tabulation the ratio of these stress values is also included.

TABLE 3

$\Delta x/L$		0,01	0,02	0,04	0,06	0,08	0,10
$\tau^{(i)}/q$	right	3,527	1,244	0,438	0,237	0,153	0,109
	left	0,0175	0,0123	0,0086	0,0069	0,0059	0,0052
Ratio		201,0	101,0	51,0	34,3	26,0	21,0

$(\Delta L/L) = 0,01$

In the example discussed, the distribution of total stresses is always symmetric with respect to the current centre line of the excavation. It is obvious from Table 3 however, that this is not the case with the induced stresses. In the vicinity of the advancing face the stress changes are very much greater than in the neighbourhood of the face opposite. This is so because the stress field around the mined face must advance with each enlargement of the excavation. This mobility of the stress field in the active region means that each mining step exposes a fresh volume of rock to stress disturbances, with a corresponding increase in the probability of encountering pre-existing weaknesses which may become the foci of new seismic events.

In summary, a convenient control of the risk of seismic events might be based on the management of stress concentration at the faces which are to be mined.

Rockbursts as a sub-set of seismic events

Virtually no systematic research has been done to elucidate the basis of setting apart those seismic events which become rockbursts from those which do not. Thus, the brief discussion presented here is based only on reasoning and not on the evaluation of objective measurements.

Rockbursts by definition are those seismic events which injure men and/or damage mining cavities or equipment in them. It is reasonable to suppose that the following factors play a major role in determining whether or not a particular seismic event becomes a rockburst:
(i) the kinetik energy content of the event,
(ii) the distance from the focus of the event to the mining excavation,
(iii) the state of stress around the excavation,
(iv) the state and quality of the rock surrounding the excavation and
(v) the quality of the support in the excavation.

It is reasonably obvious that a seismic event is more likely to have damaging effect if its energy content is high and if it occurs close to the mine opening. It is more difficult to quantify the influence of the state of stress, rock quality and the effectiveness of support.

The waves generated by a seismic event when they reach an opening cause vibration and fluctuation in stress levels. If the state of stress is favourable, that is predominantly compressive, if the surrounding rock is not too fragmented and if the support is able to knit the fractured rock together even under shock loading, then it is less likely that a seismic event will have harmful effect - hence it is less probable that it will be recorded as a rockburst.

ALLEVIATION OF THE ROCKBURST HAZARD

Summary of basic principles

The alleviation of the risks associated with rockbursts may be achieved by seeking:
(i) a reduction in the number of seismic events,
(ii) a lessening of the kinetic energy content of seismic events,
(iii) a decrease in the proportion of seismic events which manifests themselves as rockbursts, and by
(iv) minimizing the damaging effects of rockbursts.

Amongst the improvements listed here clearly some are more important than others. The fight against the harmful consequences of rockbursts, for example, through the introduction of better mine support, must go on almost regardless of any other activity which might be pursued simultaneously. Also, significant success under (i) may not turn out to be a meaningful gain. As mentioned earlier, rockbursts are only a small sub-set of seismic events. Thus, even a relatively small change in the mechanism which determines the size of this sub-set could cancel the benefits expected to arise from a decrease in the number of seismic events. This possibility, however remote it appears to be, cannot be neglected.

To achieve a reduction in risk through the first two improvements listed above it is necessary to study the criteria for the occurrence of seismic events. The utilization of the last two improvements is severely limited by various factors which will be discussed later.

Theoretical foundation

The matters discussed in this section are relevant to the control of those seismic events which occur in the vicinity of mining excavations and come about largely as a result of mining activities.

It will be recalled that for a seismic event to occur there must be points in the rock where the equilibrium may become unstable and at least one unstable equilibrium must be triggered off causing a seismic event. For this event to become of practical significance there must be a sizeable sudden change in stress at the focus and a substantial amount of energy must be stored in the region to provide the seismic energy for the event.

These criteria are obviously and directly influenced by the control of the stress concentration around the edges of tabular excavations, especially at those faces which are being worked. In the present context, a reduction in stress concentration means a lowering in the magnitude of stresses and a decrease in the volume of rock which is exposed to high stresses. If such reductions are achieved then all four conditions mentioned whould change for the better, with a consequential reduction in the number and energy content of seismic events. It must be emphasized that this change is an expectation more in a statistical than in a deterministic sense.

To implement the ideas expounded here it is necessary to find a convenient measure of energy concentration. Such a measure emerges from the earlier discussed analysis of energy changes due to mining. If mining progresses in small steps then the energy released during each advance equals the strain energy stored in the volume to be extracted, see (16).

As this paper deals with tabular deposits the formula defining this stored energy is given in (22):

$$\Delta W_r = \Delta U_m = \tfrac{1}{2} \int_{\Delta A} T_i^{(p)} \Delta s_i dA. \qquad (22)$$

An alternative expression of ΔU_m can be obtained from the strain energy density function Γ, which, in terms of principal stresses, takes the following form:

$$\Gamma = \frac{1}{2E}\{\tau_1^2 + \tau_2^2 + \tau_3^2 - 2\nu(\tau_1\tau_2 + \tau_1\tau_2 + \tau_1\tau_3)\} \quad (33)$$

Where $\tau_1 \geqq \tau_2 \geqq \tau_3$. At the face to be advanced one of the principal stresses is zero, say $\tau_3 = 0$, and the other, say τ_2, can be expressed as follows:

$$\tau_2 = \phi\tau_1, \qquad \phi \leq 1. \qquad (34)$$

The substitution of these values into (33) yields:

$$\Gamma = \frac{\tau_1^2}{2E}(1 + \phi^2 - 2\nu\phi) \qquad (35)$$

Now ΔU_m in (22) can be reformulated:

$$\Delta U_m \cong \Gamma\Delta V = \Gamma h\Delta A, \qquad (36)$$

Here $V = h\Delta A$ is the volume of rock to be mined, h is the stoping width and ΔA is area in the plane of the reef which is to be extracted in one step.

$$\Delta U_m \cong \frac{\tau_1^2}{2E}(1 - 2\nu\phi + \phi^2)h\Delta A \qquad (37)$$

If the value of ΔU_m can be obtained independently, for example from (22), then (37) can be used to calculate an estimate of τ_1:

$$\tau_1 = \sqrt{\frac{2E}{(1-2\nu\phi+\phi^2)h} \frac{\Delta U_m}{\Delta A}} \qquad (38)$$

It is noteworthy that in a two-dimensional problem where the plane strain approximation is acceptable, the value of ϕ is defined and it equals to the Poisson's ratio, that is,

$$\phi = \nu \qquad (39)$$

It is simple to prove that function $(1 - 2\nu\phi + \phi^2)$ has a minimum at $\phi = \nu$, that is at the same value which applies to plane strain problems. Thus an upper-bound of τ_1 is as follows:

$$\tau_1 \leqq \sqrt{\frac{2E}{(1-\nu^2)h} \frac{\Delta U_m}{\Delta A}}, \qquad (40)$$

where the equal sign applies in the state of plane strain. The formula in (40) is one of the most important theoretical tools in the campaign aimed at the abatement of the rockburst hazard.

It should be noted that the ratio $(\Delta U_m/\Delta A)$ is identical to the 'energy release rate' (E.R.R.) which has been used since the mid-1960's to guide mine layout design.[1] The acceptance of E.R.R. as the basis for ranking layouts was based on an erroneous reasoning. It was supposed that if the released energy is reduced then somehow seismic events will tend to be starved of energy. It has been shown earlier that this argument does not apply if mining takes place in small steps.

It is fortunate and important to observe that rankings prepared on the basis of τ_1 in (40) or on the basis of E.R.R. will be identical as long as the stoping width and rock properties remain unchanged.

It must be conceded that the stresses computed from (40) are fictitious quantities. Rock in the immediate vicinity of tabular excavations in deep mines is always fractured and as a result, de-stressed to lesser or greater extent. Nevertheless, the theoretical quantities computed from (40) can be expected to give good indications of relative merits of layouts so they might be used with reasonable confidence to handle practical problems. Wide-scale observations lend strong support to this recommendation.

Search for means of controlling stress concentration

An alternative formula of ΔU_m in (22) has been obtained recently.[30] To obtain this result first expressions are derived for the total energies released during mining from the virgin state to two neighbouring states. The difference between these energy quantities is $\Delta W_r = \Delta U_m$. This change in released energy is illustrated by the cross-hatched area in Fig. 10 which corresponds to the contribution of a small dA area of the excavation. To derive $\Delta W_r = \Delta U_m$ the contribution $d\Delta U_m$ must be integrated for the whole mined-out region. The result is reproduced here without proof and applied to the general case where a part of the tabular excavation is supported by some type of backfill:

28

$$\Delta U_m = \tfrac{1}{2} \{ \int_A Q_i \Delta s_i dA + \int_B \Delta R_i s_i^{(p)} dA - \int_{B_0} R_i^{(p)} \Delta s_i dA \}, \quad (41)$$

where Δs_i and ΔR_i denote the induced relative displacement of the hanging and footwalls and the induced change in fill resistance. Moreover, let q_{ij} be the virgin stress tensor in the plane of the reef and μ_j the interior normal of the footwall. In terms of these notations the virgin traction vector in the reef plane is as follows:

$$Q_i = \mu_j q_{ij}. \quad (42)$$

While the formula in (41) is not convenient for the numerical evaluation of ΔU_m it does provide an opportunity of showing clearly which parameter plays an important role in influencing the magnitude of ΔU_m. To this end note that $\Delta R_i = m_i \Delta s_i$ therefore (41) can be put into a slightly different form:

$$\Delta U_m = \tfrac{1}{2} \{ \int_A Q_i \Delta s_i dA + \int_B m_i s_i^{(p)} \Delta s_i dA - \int_{B_0} R_i^{(p)} \Delta s_i dA \} \quad (43)$$

This expression reveals that the most obvious way of reducing ΔU_m and, therefore decreasing the stress concentration ahead of the face which is to be advanced, (see the relationship in (40) is to minimize the induced relative displacement of the hanging and footwalls. By keeping the magnitude of $s_i^{(i)} = \Delta s_i$ as small everywhere in area A as practicable, the risk of seismic events is curbed as much as possible.

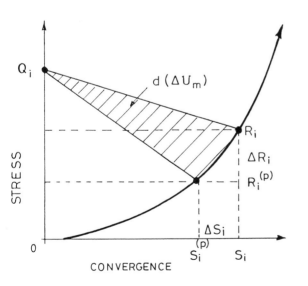

Fig. 10 Increase in released energy per unit area of the mined-out region due to some additional mining.

It should be noted that the formula in (40) can be used to estimate stress concentration anywhere at the edges of tabular excavations. If at the point where stress concentration is to be calculated no mining is planned, a fictitious face advance will provide the necessary data to permit the computation.

Strategic measures

Currently employed strategic measures of protection from seismic hazard are all based on the hypothesis that the seismicity of a region diminishes if the magnitude of the areal rate of energy release is reduced. Fortunately the practical consequences of this hypothesis are not different from those which follow from the recommendation that stress concentrations should be kept as low as possible. Perhaps the best guide for the implementation of this recommendation is that embodied in (43).

In essence three different approaches are used to achieve a reduction in stress concentration or in $\Delta U_m/\Delta A$. Of course, it must be kept in mind that $\Delta U_m/\Delta A$ must be controlled not only at any one step during mining but throughout the extraction of the whole area in question. The control measures applied are as follows:

(i) planning of the layout and the sequence of mining,
(ii) reduction in stoping width,
(iii) backfilling, and
(iv) partial extraction.

Layout planning is based on the notion that virtually on all occasions a given area can be mined several ways. The ultimate choice should be based on the analysis and ranking of the various options on the basis the ratio $\Delta U_m/\Delta A$ or the areal rate. (Note: In practice it is preferable to use $\Delta U_m/\Delta A$ instead of the values of stress concentration because the latter involves the square root of the areal rate, thus it is necessarily less sensitive, see (40)). The object of the planning is to ensure that the areal rate remains as uniform as practicable throughout the mining of the area, avoiding geometries which temporarily retard the development of convergence and ride components. Remnants, pillars, etc. initially hinder the growth of the relative displacement components,

but when they themselves are being extracted
there is an acceleration in convergence and ride
with a commensurate increase in the value of
$\Delta U_m/\Delta A$.

Layout planning is performed with the aid of
electric analogues[23] or digital systems.[24,25]
All deep mines of the gold mining industry apply
either of these methods in mine planning on a
systematic basis.

Reduction in stoping width can be considered
only in situations where the current working width
is considerably greater than both the acceptable
minimum working width (say 1m) and the width of
the gold-bearing channel. In instances where
these criteria are satisfied there is considerable
incentive, both economic and rock mechanics, to
lower the stoping width to the tolerable minimum.

Backfilling has two advantages. Firstly,
it reduces the width which can be taken up by
convergence and this lessening effect applies
both to the ultimate convergence volume and to
the induced displacement at each step in mining.
Secondly, the rock mass must do some work on
the backfill to compact it. In the process
some energy is consumed. As a result the energy,
which otherwise would have been available to
contribute to $\Delta U_m/\Delta A$, is reduced. In fact, the
negative terms in (41) and in (43) represent the
quantum of energy which is expended in compressing
the fill.

Some years ago Ryder and Wagner examined[39]
the implications of employing backfill and found
that the compression characteristics of possible
fill materials can be described surprisingly
well by

$$\sigma = \frac{a\varepsilon}{b - \varepsilon}, \qquad (44)$$

where a and b material constants and ε is the
compressive strain. Let the stoping width be h,
the uncompacted width of the fill at installation
be w and the unresisted convergence be s_0 that is
$h = w+s_0$. In terms of these notations

$$\varepsilon = \frac{s-s_0}{w}, \qquad (45)$$

therefore the compression characteristics takes
the form

$$\sigma = \frac{a(s-s_0)}{bw-(s-s_0)} \qquad (46)$$

Here s is total convergence.

It is possible to evaluate analytically and
rank the performance of fill materials for the
limiting case when a long face is advanced and
the span is assumed to be infinite. In this
idealized example the energy balance of a small
face advance can be computed by assuming that an
equivalent area is added to the region where the
pressure on the fill has reached the ultimate
value, that is the virgin vertical stress.
Thus, in this region $\sigma= Q$ which condition
permits the computation of the maximum convergence
s_m, from (46),

$$\frac{s_m}{h} = 1 - (1 - \frac{b\beta}{1+\beta})\alpha, \qquad (47)$$

where the following notations are employed:

$$\alpha = \frac{w}{h}, \qquad \beta = \frac{Q}{a}. \qquad (48)$$

The ratio α can be referred to as the 'lag
factor' because it is a measure of the delay in
the placing of fill. Parameters α and β are both
dimensionless.

The work done in compacting the fill over a
unit area is defined by the following integral:

$$\frac{dW_s}{dA} = \int_{s_0}^{s_m} \sigma(s)ds = abw\{ln(1+\beta)- \frac{\beta}{1+}\} \qquad (49)$$

The relationship in (23) yields

$\Delta U_m = \Delta W - \Delta(U-U') - \Delta W_s$, but in this example
$U = U'$ therefore

$$\frac{dU_m}{dA} = \frac{dW}{dA} - \frac{dW_s}{dA}. \qquad (50)$$

It is obvious from this result that without
backfill

$$\frac{dU_m}{dA}\Big|_{max} = \frac{dW}{dA}\Big|_{max} = Qh \qquad (51)$$

If, however, backfill is used the work done by
the body forces becomes

$$\frac{dW}{dA} = Qs_m \qquad (52)$$

and the effectiveness of the fill can be measured
by the following ratio:

$$\frac{dU_m}{(dU_m)_{max}} \frac{dW-dW_s}{(dU_m)_{max}} = 1-\alpha\{1-b[1-\frac{1}{\beta}ln(1+\beta)]\} \qquad (53)$$

30

Ryder and Wagner suggested that a 'factor of merit', M, can be defined for a fill material. This is obtained from (53) by putting the lag factor $\alpha = 1$:

$$M = b\{1 - \frac{1}{\beta} \ln(1+\beta)\} \qquad (54)$$

The <u>lower</u> the value of M the effect of the fill becomes <u>more</u> beneficial. In Table 4 the factors of merit are given

Table 4 Factor of merit for backfill[39]

Material	a MPa	b	Depth m		
			1 000	2 000	3 000
A	20,57	0,214	0,077	0,109	0,127
B	6,52	0,368	0,223	0,269	0,291
C	1,30	0,547	0,466	0,498	0,511

$Q = 0,027$ Depth (MPa)

for three materials when applied at various depths. It seems that the factor is affected relatively little by depth and, therefore, it can be regarded largely the property of the material. This conclusion applies especially if the depth of mining exceeds 2 000 m.

According to the factors of merit in Table 4 materials A, B and C would furnish progressively less effective fill. In Fig. 11 the compression characteristics of these materials are depicted. Obviously, substance A is the 'stiffest' and C is the 'softest' fill, indicating the need that the aim should be to place as stiff a fill as practicable to achieve the best result. Furthermore, it is important to ensure that the placing of the fill is done as close to the face as possible and it keeps pace systematically with the advance of mining. Much of the advantages of backfilling can be wasted if this is not achieved. A quick substitution into (53) proves this suggestion. For example, Table 4 gives $M = 0,109$ for material A at the depth of 2 000 m. According to (53), this value increases to 0,198 if placing is delayed so as to have $\alpha = 0,9$. This change represents some 35 per cent increase in stress concentration.

The potential improvements resulting from the application of backfill were recognized already in the 1960's.[1] However, practical problems have so far prevented the introduction on a significant scale of this method of rock-burst control. The difficulties are mainly associated with the logistic of transporting the extremely abrasive fill material and placing it in the narrow stopes in a manner which does not interfere with the normal stoping operations. In recent years, however, a determined effort is being concentrated on the development of filling methods which are suitable in our deep mines. It is confidently predicted that the introduction of backfilling in some form is imminent.

<u>Partial extraction</u> is the most aggressive method of combating the rockburst hazard. The concept is to design a regular layout of permanent stabilizing pillars with the view of reducing the volume convergence considerably.[1] The kernel of this idea is to ensure that the pillars are able to support the overburden. This is achieved by ensuring that the width-to-height ratio of pillars is large, say in excess of 20.

The most obvious layout of partial extraction is based on parallel sided long panels separated by strip pillars, where mining proceeds along the long axis of the system. It has been shown that this layout leads to a considerable reduction in stress concentration.[1] Perhaps the most significant advantage of this type of partial extraction is that the stress concentration remains virtually constant as soon as the length of the panels becomes two or three times greater than their span.

Several of the deepest mines of the industry have introduced versions of this partial extraction with impressive results in terms of reduced seismicity.[34] It is noteworthy that this was done in spite of the significant, say 15 per cent, loss of revenue resulting from the sacrifice of the gold which is permanently locked up in the pillars.

There is an apparent weakness in this layout. If the extraction ratio is high then the stress concentration at the edges of pillars is considerably higher than at the moving faces. This suggests that seismic activity might be expected to occur in the vicinity of these edges. Fortunately experience shows no or very limited seismic activity in this region. This feature of the system must be attributed to the almost total absence of triggering stress changes. Nevertheless, the possibility of an occasional occurrence of seismic events near the edges cannot be excluded, therefore travelling ways must be sited far from the highly stressed pillars.

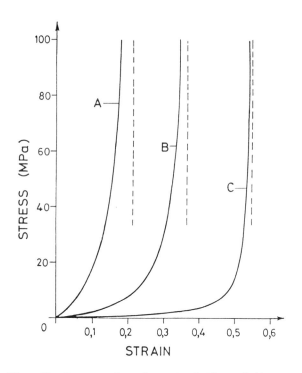

Fig. 11 Compression characteristics of three fill materials

Partial extraction can be employed in conjunction with backfilling also. Gold recovery can be improved by widening the spacing of pillars and introducing fill into the centre region of the wider panels. Conceptually it is also possible to replace the pillars with ribs constructed of backfill. The benefit accruing from these methods are obviously less than those attributable to the system based on solid pillars, so it should be used only in situations where the rockburst problem is less acute.

Tactical measures
Regretfully it must be conceded that no strategic measure is likely to lead to the total elimination of seismic events. Also, at the present depths of gold mining every cavity is surrounded by fractured rock. In view of combined effects of these observations, it is obvious that effective support systems are required to minimize the chances that a seismic event causes the voilent collapse of nearby excavations.

The need for efficient support was recognized many years ago. However, only after the advent of the seismic investigations did it become possible to formulate specifications for a support system which would remain effective even when subjected to shock loads. N.G.W. Cook did in fact formulate such specifications on the basis of his seismic research and his ideas did eventually lead to the construction and proving of the 'rapid yielding hydraulic props'.[40]

During about the same period conventional stope support was also undergoing some fundamental changes. It became obvious that the all-timber packs were able to provide significant support resistance only after suffering considerable deformation. Thus, they provided little if any support in the vicinity of the stope face where men were most exposed to hazards. The stiffness of these packs was enhanced considerably through the introduction of pre-fabricated concrete members into their structure.[41,42,43]

These early improvements gave impetus to the development of new support systems based on some combination of timber props and steel pipes. The idea was to enhance the strength of the props and at the same time incorporate facilities for some controlled yield.

In summary, the new ideas of stope support involve increased stiffness and the property of controlled rapid yield. Also, it has become accepted that an ideal support should be active, that is, it should provide resistance to convergence immediately on installation. The hydraulic prop and the more recently developed rapid yielding hydraulic chocks incorporate all of these concepts.

Design and support of tunnels and service excavations
Most of the discussion so far has been devoted to stopes. Of course, rockbursts endanger also all tunnels, shafts and service excavations. Thus, their siting and support require considerable attention. Unfortunately, it is not possible to discuss here the developments which have taken place in this field of rock mechanics. However, the relevant material is presented in some detail in a recent report which is available for study.[2]

FUTURE RESEARCH

Considerable progress has been made during the last three decades in the field of rock mechanics as related to the problem of rockbursts. Nevertheless, many questions remain unanswered. An attempt is made in this section to highlight some of these outstanding problem areas.

The possibility of predicting rockbursts receive tremendous attention from miners and from the press. This task, because of its obvious importance, has received considerable attention from researchers all over the world, including South Africa. With advances in science and technology more and more complex attacks are made on the problem. The current idea is to explore the possibility of studying the distribution of seismicity, both in time and in space using an advanced seismic network enhanced by on-line computers to facilitate analysis in real time. The problem is a very difficult one. It is not possible to predict whether or not the present attempt will be successful.

Earlier in discussing seismic events, qualitative ideas, (based on pre-existing weaknesses, on the magnitude of stresses and of triggering stress changes), were put forward to explain the underlying mechanism. It seems possible that these ideas can be formalized into a quantitative model with the facility of yielding statistical estimates of the probability of seismic events occurring in a given situation.

It is clear that geological weaknesses, such as joints, faults and dykes, play an important role in the initiation of seismic events. Numerical techniques are now available to facilitate the quantitative analysis of this type of problem.[44,45] It can be predicted that significant progress in the solution of this problem is imminent.

As mentioned earlier, virtually no systematic attention has been given to the factors determining which of the seismic events will cause damage to mining excavations. It is likely that a detailed study of this question, using seismic and other means of observation, would yield conclusions which could be employed to improve the safety of miners and the integrity of cavities.

At one time it was thought that the rockburst hazard could be alleviated by creating a 'de-stressed' region ahead of the face with the aid of heavy blasting in long holes. Detailed seismic studies indicated that no excess of energy, other than that which could have been derived from the explosive, was radiated as a result of either de-stressing or conventional blasts.[1] Perhaps the new ideas about instability and triggering stress changes might give sufficient prompting to give further attention to blasts which might provoke seismic events under controlled conditions.

The seismicity around Klerksdorp appears to have significantly different features from those observed elsewhere. It has been suggested that in this area tectonic forces may have brought geological features to the brink of unstable equilibrium. If this is in fact so then minor stress changes induced by mining may trigger off seismic events. Considerable evidence has been accumulated to support this supposition. It would be important to decide whether the idea has factual basis and if so, how the geological features on the brink of instability could be identified. Also, if the identification is successful, would it be possible to trigger off the instability in a safe manner, for example, by 'lubricating' the surfaces of the appropriate geological feature using water injection?

Perhaps the greatest challenge to solid state physics today is the modelling of the behaviour of strain-softening medium. Considerable attention is being concentrated on this problem, so presumably a solution will emerge. But the weight of the problem is not to be underestimated so the efforts must be sustained.

There are many other problems which could have been included here. It is hoped that the selected examples will be sufficient to show the magnitude of the challenge which faces the researchers wishing to tackle the rockburst problem.

CONCLUSION

This paper was intended to do two things. Firstly, to review three decades of research aimed at the solution of the rockburst problem and secondly, to explain, in the light of more recent theoretical results, the present understanding of the mechanism underlying the phenomenon.

The tasks have proved to be formidable and commitments competing for attention did not help. Consequently, the final product turned out to be longer in length but lighter in content than it was originally envisaged.

References
1.   Cook N.G.W., Hoek E., Pretorius J.P.G., Ortlepp W.D. and Salamon M.D.G. Rock mechanics applied to the study of rockbursts. Journal of the South African Institute of Mining and Metallurgy, vol. 66, 1966, p. 435-528.
2.   An industry guide to the amelioration of the hazards of rockbursts and rockfalls. Compiled by the High Level Committee on Rockbursts and Rockfalls, Research Organization, Chamber of Mines of South Africa, 1977, 178 p.
3.   Morris D and Vorster F. A record of fall-of-ground accidents on gold mines. Seismicity in Mines, South African National Group for Rock Mechanics. Preprints: Session 4, Johannesburg, September, 1982.
4.   Hill F.G. A system of longwall stoping in a deep level mine, with special reference to its bearing on pressure bursts and ventilation problems. Association of Mine Managers of the Transvaal, Papers and Discussions, 1942/45, vol. I, January, 1944, p. 257-276.
5.   Hoek E. and Bieniawski Z.T. Brittle fracture propagation in rock under compression. International Journal of Fracture Mechanics, vol. 1, 1965, p. 137-155.
6.   Hoek E. and Bieniawski Z.T. Fracture propagation mechanism in hard rock. Proceedings of the 1st Congress of the International Society of Rock Mechanics, Lisbon, vol. 1, 1966, p. 243-249.
7.   Cook N.G.W. The failure of rock. International Journal of Rock Mechanics and Mining Sciences, vol. 2, 1965, p. 389-403
8.   Deist F.H. A non-linear continuum approach to the problem of fracture zones and rockbursts. Journal of the South African Institute of Mining and Metallurgy, vol. 65, 1965, p. 502-522

9.   Cook N.G.W. The seismic location of rockbursts. Proceedings of the 5th Rock Mechanics Symposium, Oxford, Pergamon Press, 1963, p. 493-516.
10.   Cook N.G.W. The application of seismic techniques to problems in rock mechanics. International Journal of Rock Mechanics and Mining Sciences, vol. 1, 1964, p. 169-179.
11.   Salamon M.D.G. Elastic analysis of displacements and stresses induced by mining of seam or reef deposits - Part I : Fundamental principles and basic solutions as derived from idealized models. Journal of the South African Institute of Mining and Metallurgy, vol. 63, 1963, p. 128-149.
12.   Salamon M.D.G. Elastic analysis of displacements and stresses induced by mining of seam or reef deposits - Part II : Practical methods of determining displacement, strain and stress components from given mining geometry. Journal of the South African Institute of Mining and Metallurgy, vol. 64, 1964, p. 197-218.
13.   Salamon M.D.G. Elastic analysis of displacements and stresses induced by mining of seam or reef deposits - Part IV : Inclined reef. Journal of the South African Mining and Metallurgy vol. 65, 1964, p. 319-338.
14.   Hill F.G. An investigation into the problem of rockbursts, an operational research project. Part I : The approach to the problem and analyses of the rockbursts that have occurred on the E.R.P.M. during the years 1948-1953. Journal of the Chemical Metallurgical and Mining Society of South Africa, vol. 55, October, 1954, p. 63-83.
15.   Cook N.G.W. Seismicity associated with mining. 1st International Symposium on Induced Seismicity. Canada, September, 1975
16.   McGarr A. Some applications of seismic source mechanism to assessing underground hazard. Seismicity in Mines, South African National Group for Rock Mechanics, Preprints : Session 4, Johannesburg, September, 1982.
17.   Gay N.G., Spencer D., van Wyk J.J. and van der Heever P.K. The control of geological an mining parameters on seismicity in the Klerksdorp gold mining district. Seismicity in Mines, South African National Group for Rock Mechanics, Preprints : Session 3, Johannesburg, September, 1982.

18. Salamon M.D.G. Contribution to the paper : 'Rockburst phenomena in the gold mines of the Witwatersrand : a review,' by J.C. Curtis. Institution of Mining and Metallurgy. Transactions, Section A, vol. 90. October, 1981, p. A212-A213.

19. Hackett P. An elastic analysis of rock movements caused by mining. Transactions of the Institution of Mining Engineers, vol. 118, Part 7, 1959, p.421-435.

20. Salamon M.D.G. An introductory mathematical analysis of the movements and stresses induced by mining in stratified rocks. King's College, University of Durham. Department of Mining. Research Report, vol. 9, Bulletin No. 3, Series: Strata Control/Res. No. 16, 1961.

21. Salamon M.D.G. The influence of strata movement and control on mining development and design. Ph. D. Thesis, University of Durham, 1962.

22. Salamon M.D.G. Rock mechanics of underground excavations. Advances in Rock Mechanics : Proceedings of the 3rd Congress, International Society for Rock Mechanics, Denver, Colorado, vol. 1, Part B, p. 951-1099. Washington, National Academy of Sciences, 1974.

23. Salamon M.D.G., Ryder J.A. and Ortlepp W.D. An analogue solution for determining the elastic response of strata surrounding tabular mining excavations. Journal of the South African Institute of Mining and Metallurgy. Vol. 65, 1964, p. 115-137.

24. Starfield A.M. and Fairhurst C. How high-spread computers can advance design of practical mine pillar systems. Engineering/Mining Journal, vol. 169, 1968, p. 78-84.

25. Plewman R.P., Deist F.H. and Ortlepp W.D. The development and application of a digital computer method for the solution of strata control problems. Journal of the South African Institute of Mining and Metallurgy, vol. 70, 1969, p. 33-44.

26. Wilson J.W. and More O'Ferrall R.C. The application of the electrical resistance analogue to mining operations. Journal of the Institute of Mining and Metallurgy, vol. 70, 1970, p. 115-148.

27. Salamon M.D.G. Elastic analysis of displacements and stresses induced by mining of seam or reef deposits - Part III : An application of the elastic theory : Protection of surface installations by underground pillars. Journal of the South African Institute of Mining and Metallurgy, vol. 64, 1964, p. 468-500.

28. Wagner H and Salamon M.D.G. Strata control techniques in shafts and large excavations. Association of Mine Managers of South Africa. Papers and Discussion, 1972-1973, p. 123-140.

29. Salamon M.D.G. Energy considerations in mining - Part I : Fundamental results. Journal of the South African Institute of Mining and Metallurgy. To be submitted for publication.

30. Salamon M.D.G. Energy consideration in mining - Part II. Journal of the South African Institute of Mining and Metallurgy. In preparation.

31. Drucker D.C. On the postulate of stability of material in the mechanics of continua. Journal de Mécanique, vol. 3, No. 2 1964. p. 235-249.

32. Salamon M.D.G. Stability, instability and design of pillar workings. International Journal of Rock Mechanics and Mining Sciences, vol. 7, 1970, p. 613-631.

33. Brady B.H.G., and Brown E.T. Energy changes and stability in underground mining : design applications of boundary element methods. Institution of Mining and Metallurgy. Transactions Section A, vol. 90, April 1981, p. A61-A80.

34. Salamon M.D.G. and Wagner H. Role of stabilizing pillars in the alleviation of rockburst hazard in deep mines. Proceedings of the 4th Congress, International Society of Rock Mechanics, September, 1979, Montreaux, vol. 2, p. 561-566. Balkema, Rotterdam, 1979.

35. Bryan A., Bryan J.G. and Fouché J. Some problems of strata control and support in pillar workings. Mining Engineer, vol. 123. 1964, p. 238-254.

36. Spackeler G., Gimm W., Höfer K.H. and Duchrow G. New data on rockbursts in potash mines. Proceedings of the 3rd International Conference on Strata Control, Paris, 1960. Ed. de la Revue de l'Industrie Minerale, p. 551-563.

37. Rice J.R. Mathematical analysis in the mechanics of fracture. In 'An advanced treatise of fracture'. Ed. H. Liebowitz, vol. 2, p. 192-311, Academic Press, 1968.

38. Hodgson K. and Joughin N.C. The relationship between energy release rate, damage and seismicity in deep mines. Proceedings of the 8th Symposium on Rock Mechanics, University of Minnesota. Ed. C Fairhurst, A.I.M.E., p. 194-209, New York, 1966

39. Ryder J.A. and Wagner H. 2D analysis of backfill as means of reducing energy release rates at depth. Chamber of Mines of South Africa, Research Organization, Research Report No. 47/78, 1978, 22 p

40. Tyser J.A. and Wagner H. Review of six years of operations with the extended use of rapid-yielding hydraulic props at the East Rand Proprietary Mines, Limited and experience gained throughout the industry. Association of Mine Managers of South Africa. Papers and Discussions, 1976-1977, p. 321-347.

41. Margo E. and Bradley R.K.O. An analysis of the load compression characteristics of conventional packs. Journal of the South African Institute of Mining and Metallurgy, vol. 66, 1966, p. 364-401.

42. Cook N.G.W. Contribution to the paper 'An analysis of the load compression characteristics of conventional packs' by E Margo and R.K.O. Bradley, Journal of the South African Institute of Mining and Metallurgy, vol. 66, 1966, p. 643-648.

43. Joughin N.C. The reduction of stope damage resulting from the use of concrete for support at the Harmony Gold Mining Co., Ltd. Chamber of Mines of South Africa, Research Organization, Research Report No. 69/67, 1967.

44. Ryder J.A., Ling T. and Wagner H. Slippage along a fault plane as a rockburst mechanism - static analysis. Chamber of Mines of South Africa, Research Organization, Research Report No. 31/78, 1978, p.17

45. Crouch S.L. Computer simulation of mining in faulted ground. Journal of the South African Institute of Mining and Metallurgy, vol. 79, 1979, p. 159-173.

Regional aspects of mining-induced seismicity: theoretical and management considerations

E.L. Dempster A.C.S.M., C.Eng., F.I.M.M.
Research Organization, Chamber of Mines of South Africa, Johannesburg, South Africa
J.A. Tyser B.Sc.(Min.), Pr. Eng., F.S.A.I.M.M.
Blyvooruitzicht Gold Mine, Transvaal, South Africa
H. Wagner Dipl. Eng. (Min.), M.S.A.I.M.M.
President Steyn Gold Mine, Orange Free State, South Africa

SYNOPSIS

The paper discusses differences in seismicity in the major gold mining districts. It is shown that on a regional scale seismicity is related to the mining induced energy changes but that the pattern of seismicity is governed by geological structures. The differences in the nature of the rockburst problem are discussed on the example of two major gold mines. In the case of Blyvooruitzicht gold mine which operates in a relatively unfaulted mining district the rockburst hazard is closely related to the stoping geometry while at the Vaal Reefs gold mining complex the rockburst problem is closely linked to the geological structure and affects both on-reef as well as off-reef excavations. Examples of how mine management deals with the rockburst problem on these two mines are given.

INTRODUCTION

The occurrence of sudden rock failures which radiate seismic energy is one of the most serious problems associated with the extraction of many of the gold bearing reefs of the Witwatersrand system. A small fraction of these mining induced seismic events damage underground excavations and, occasionally, surface structures, and cause accidents which often result in production losses. These events are known as rockbursts and have been the subject of many investigations.

The purpose of this paper is to show that significant regional differences exist in the nature and magnitude of rockburst problems and that mining induced seismicity can have many facets. It is for these reasons that rockburst control measures and strategies differ from area to area. This is illustrated by citing examples from two major gold mines which are situated in the Far West Rand and Klerksdorp goldfields.

DIFFERENCES IN SEISMICITY IN THE VARIOUS GOLD MINING DISTRICTS

Mining of gold bearing reefs takes place on the periphery of the Greater Witwatersrand basin which stretches 320 km in a north-easterly direction and 160 km in a north-westerly direction (Fig. 1). Extensive major faulting has taken place along the edge of the basin, parallel to and transverse to the strike of the beds. In many instances the reefs have been displaced by numerous relatively minor strike and transverse faults. These are mostly normal faults, but reverse and strike-slip faults can also be encountered. Dykes, which occupy many of the faults cutting the reefs, are generally of a basic composition, but syenitic and more acidic varieties are also present. Their ages vary from Ventersdorp through Pilanesberg and Karoo to post-Cretaceous.

At present significant gold mining operations take place in the Evander area on the eastern extremity of the Greater Witwatersrand basin, the East Rand, the West Rand and the Far West Rand, the Klerksdorp district and in the Orange Free State goldfield. Fig. 2 shows that there are significant differences in the quarterly production figures of the various gold mining districts but that the variation in production figures during the past decade was remarkably small, with the exception of the old mining areas on the East and Central Rand. Fig. 3 gives an indication of the differences in average rock

breaking depths between the various gold mining
districts. The trend reflected in this figure is
influenced strongly by the closing of old and
opening of new mines. Except for the mines in
the Evander area and the old mines on the East
Rand, mining takes place at an average depth of
between 1600 m and 2000 m. However, in individual
districts the depth of mining can vary from 1000 m
to more than 3000 m.

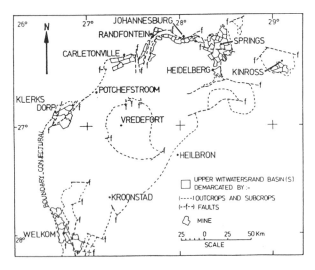

Fig. 1 Sketch map showing the periphery of the
upper Witwatersrand basin(s), major faults and
mines[5]

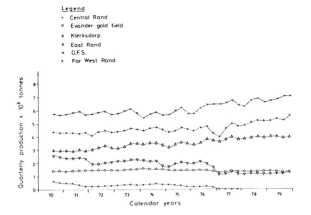

Fig. 2 Quarterly production figures for the
major gold mining districts (1970-1980)

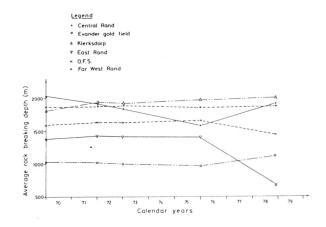

Fig. 3 Variation in average rock breaking depth
for the major gold mining districts (1970-1978)

Using seismic data from the national seismic
network which is operated by the Geological Survey
of South Africa, it was possible to construct
seismic histories for the various gold mining
districts for the period 1970 to 1980.[5] Because
of the large distances between individual geophone
stations the location accuracy of individual
seismic events is of the order of 2 km to 10 km,
which is often insufficient to allocate a seismic
event to an individual mine, but generally
accurate enough to pinpoint the mining district
where the event originated from. For similar
reasons only events with Richter magnitudes in
excess of $M_L \geq 1,9$ have been recorded. Fig. 4
shows a plot of the cumulative seismic energy
for the major gold mining districts. According
to Fig. 4 the cumulative seismic energy which was
observed in the Far West Rand goldfield was more
than twice as high as that recorded in the next
most active districts, namely the Klerksdorp and
Orange Free State goldfields. It is also noteworthy that the level of seismicity in the
shallowest of the gold mining districts, the
Evander goldfield, was only 1,5 per cent of that
of the Far West Rand. However, a direct comparison of the seismic activities in the various
districts is misleading, as there are substantial
differences in both the amount of rock mined as
well as the average rock breaking depths.

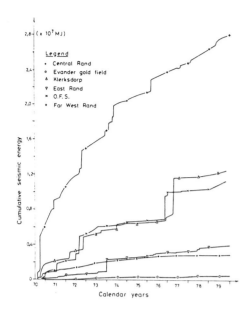

Fig. 4 Cumulative seismic energy for major gold mining districts (1970-1980)

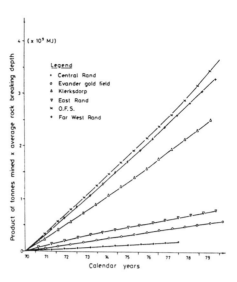

Fig. 5 Cumulative potential energy characteristics for major gold mining districts (1970-1980)

These differences have been accounted for in Fig. 5 which shows the cumulative product of the tonnes mined and the average rock breaking depth. This product is not only an excellent measure of the mining effort, but it also provides an indication of the potential energy changes caused by the extraction of the gold bearing reefs. It is important to note that in the case of the enlargement of excavations in small steps the change in potential energy is fully expended in straining of the rockmass and the deforming of the supports. Salamon[6] has shown that in the case of very extensive mining and a constant stoping width, the potential energy changes due to advancing a stope face in small steps are given by the product of stoping width and virgin vertical rock stress. Since the tonnes mined and stoping width are related by the density of the rock and the virgin vertical rock stress is proportional to depth, it follows that the product of tonnes mined and average rock breaking depth can be employed to compare mining induced energy changes in the various districts. In doing so it must be remembered, however, that the mining geometry and in particular the percentage extraction have a strong influence on the magnitude of the potential energy changes. It is for this reason that the term potential energy characteristic (P.E.C.) has been chosen to describe the product of tonnes mined and average rock breaking depth.

A comparison of the cumulative potential energy characteristics of the various mining districts with the cumulative seismic energy is given in Table 1. The seismic characteristic of a district is defined as the ratio of the seismic energy and the potential energy characteristic multiplied by hundred. Table 1 shows that the seismic characteristic of the Central Rand is highest at 1,35, followed by the Far West Rand at 0,84. The East Rand and Klerksdorp districts and the O.F.S. goldfield have very similar seismic characteristics, while the level of seismicity in the Evander area is extremely low. The very high level of seismicity on the Central Rand can be explained by the great depth of mining and the fact that because all mines in the district were close to the end of their active lives, mining activities were confined to the extraction of remnant areas and shaft pillars.

Table 1. Cumulative seismic energy, potential energy characteristic and seismic characteristic for the major South African gold mining districts (1970-1980)

District	Cumulative seismic energy SE (10^6MJ)	Potential energy characteristic PEC (10^9 MJ)	Seismic characteristic (SE/PEC). 100
Evander	0,4	0,55	0,07
East Rand	4,1	0,8	0,51
Central Rand	2,7	0,2	1,35
Far West Rand	27,8	3,3	0,84
Klerksdorp	12,8	2,6	0,49
O.F.S.	11,8	3,65	0,32

In an attempt to clarify further the differences in seismicity between various gold mining districts, the mining induced energy changes have been quantified more accurately by using the well established concept of the spatial rate of energy release which was first introduced by Cook and later substantiated by a number of researchers.[3,4] Salamon and Wagner[7] have shown that energy release rate is a good measure of the probability of triggering a seismic event since it describes both the strain energy stored in the rock mass surrounding the working faces and the magnitude of the changes caused by advancing a stope face.

Table 2 shows the breakdown of stoping activities in the various gold mining districts into ERR categories. Using this information, which was collected in 1975 on behalf of a committee on rockbursts and rockfalls, a weighted average ERR per unit area mined in MJ/m^2 has been calculated for the various gold mining districts.

Table 2 Distribution of stoping areas according to energy release rate values (1975)

District	Percentage of stoping in different energy release rate categories (MJ/m^2)					Weighted average (MJ/m^2)	Area Stoped ($10^6 m^2$/yr)
	0-20	20-40	40-60	60-80	+ 80		
Evander	80	16	4			14,8	1,6
Central Rand	15	35	40	7,5	2,5	39,75	2,7
West and Far West Rand	34	33	23	7,5	2,5	32,5	5,0
Klerksdorp	65	23	9	3	-	20	3,6
O.F.S.	57	33	9	1	-	20,8	5,5

Fig. 6 gives the relationship between the weighted average rate of energy release on a district basis and the radiated seismic energy per unit area mined. The latter figure was derived from the cumulative seismic energy over the ten-year period from 1970 to 1979 and the total area which was stoped during this period. Fig. 6 reveals a remarkably well-defined relationship between the weighted average, energy release rate and the specific seismic energy. This finding is somewhat surprising when differences in the local geology of the various mining distracts are considered. In particular the pattern of seismicity in the Klerksdorp and O.F.S. goldfields, both of which are extensively faulted with fault displacements in excess of several hundred metres, differs significantly from that observed in the other districts. While in most of the other districts the bulk of the seismic energy

is released by a large number of smaller events, most of the seismic energy in the Klerksdorp and O.F.S. goldfields is released in a small number of very large events. These differences will be discussed in greater detail. No satisfactory answer has been found for the almost identical seismic energy release pattern, considering that the two districts are more than 100 km apart.

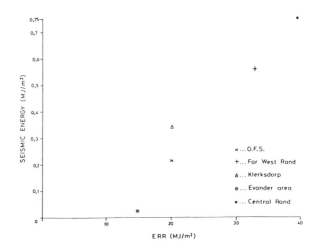

Fig. 6 Relationship between seismic events per unit area mined and weighted average ERR for the major gold mining districts.

THE ROLE OF LOCAL GEOLOGY IN THE CHOICE OF MINING METHOD AND SEISMICITY

The effect of local geology on the choice of mining method and seismicity will be discussed with the aid of examples of the Blyvooruitzicht Gold Mine which is situated on the Far West Rand some 80 km west of Johannesburg, and the Vaal Reefs gold mining complex in the Klerksdorp gold mining district. Geological conditions in these two districts differ greatly as does the general pattern of seismicity.

BLYVOORUITZICHT GOLD MINE

At Blyvooruitzicht Gold Mine Limited the Carbon Leader reef is being extracted at a depth ranging from 1400 m to 2380 m below surface. The narrow Carbon Leader reef dips at an angle of 21 degrees towards south and is very consistent on strike and dip. Faulting, although present, does not create a serious problem as far as the mine layout and choice of stoping system are concerned. Fault spacing is typically of the order of 100 m and fault displacement generally less than 1 m.

Of more serious concern is a system of near vertical dykes which strike in a north-easterly-south-westerly direction. A second but less dominant set of dykes strikes in the north-south direction.

Until 1966 mining at Blyvooruitzicht was restricted to the upper half of the mine above 16 level. The layout employed was the scattered mining or commonly termed Goldfields system whereby raises were developed from crosscuts spaced some 200 m apart on strike. This method resulted in island remnants being left between these raises which proved difficult and often impossible to extract and were frequently the site of serious rockbursts. Also, the need to develop footwall haulages ahead of stoping and to overstope them at a later stage resulted in serious support problems in the haulages, high support costs and unsafe mining conditions.

For these reasons it was decided to adopt the longwall mining method for the extraction of the Carbon Leader below 16 level, that is a depth of 1900 m below surface.

Five inclined shaft systems spaced apart at a distance of approximately 1000 m were developed to a depth of 2400 m. Each of the inclined shafts served two longwalls which advanced away from the shaft and had an extent in the dip direction of 1400 m. All footwall development except the 30 level haulage at the bottom of the lease area was carried in stoped out, that is destressed, ground. On the stoping horizon diagonal gullies were used in place of strike gullies, making it possible to discharge the ore from the face into the gullies, without the need for either an advanced heading or a step in the longwall face every 45 m which is the length of an individual stope panel.

The thinking behind this mining method was that initially the straight face shape leads to an even energy release rate and a lower incidence of rockbursts. One of the drawbacks is a tendency for a larger length of face to be damaged simultaneously because the faces are in line. Another problem is that when the longwalls approach each other in parallel fashion, peninsular remnants would be formed which would magnify the rockburst hazard.

Two alternatives were considered to minimize problems during final stages of mining. One was to have an underhand peaked longwall with the final remnants developed on 30 level, and the other to have an overhand longwall with the final remnant being extracted close to the 16 level elevation. It was reasoned that the stresses here would be lower and therefore preferable.

During 1974 it was found to be very difficult to maintain the planned face shape on the B1 longwalls due to the presence of a major dyke system which was encountered on dip. It thus required much greater effort to significantly increase face advance in the lower portions of the mine where the dyke was encountered and to slow down the face advance in the upper portions of the mine. The mining of dykes is significantly more difficult than mining of quartzite, notwithstanding the increased rockburst hazard.

After three rockbursts in the dyke which occurred in August and September, 1977, a complete change in strategy was made. A system of mining was adopted in which the longwalls were divided by continuous strike stabilizing pillars. As a consequence, a change was also made from diagonal gullies to strike gullies.

The role of the pillars was twofold:
1) to reduce the volumetric convergence and consequently the energy release rate, and
2) to provide a 'clamping' effect to halt the propagation of any fracture along the face. This was the case particularly if a geological discontinuity was approached broadside on-

Fig. 7 shows a plan of mine layout at Blyvooruitzicht gold mine. The large number of small remnants above 16 level is typical of the scattered method of mining which was adopted in the early stages of mining. The difficulties experienced in maintaining a straight longwall face over a dip distance of more than 1400 m are clearly visible.

In an attempt to monitor the effectiveness of the stabilizing pillars and to identify areas of abnormal seismicity, a seismic network was installed in the late 1970's and seismicity has been carefully monitored since June, 1979. The seismic system was operational for 95 per cent of

the time and every seismic event with a Richter
magnitude greater than 1,5 was located with an
accuracy that varied from 30 m in the middle to
approximately 150 m towards the edge of the mine.

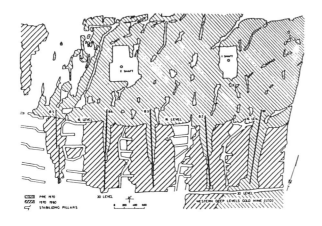

Fig. 7 Mine plan - Blyvooruitzicht Gold Mine

During most of this period potential rockburst reports, based on seismic locations, were
sent to production personnel with a request that
any rockburst damage in the stopes be indicated
on a sketch. Over 200 rockbursts of sufficient
severity to cause production losses have been
documented in this way. The return rate, which
was sometimes less than 50 per cent, did not
depend noticeably on the magnitude of the seismic
event and the incidence of reported rockbursts
was corrected for the return rate of the forms to
indicate the actual rate of rockburst occurrence.

The ERR for all mining at Blyvooruitzicht
has been routinely determined on a quarterly
basis since 1979, using convergence and stress
estimates calculated by a computer program.

Eighteen regions within Blyvooruitzicht
were considered on the basis of distinguishable
concentrations of seismicity. These regions were
then grouped into four ERR ranges, of namely,
0-20, 21-40, 41-80 and 81-160 MJ/m^2. In Fig. 8
it can be seen that the rate of seismicity,
expressed as the number of seismic events with
M>1,5 per 1000 m^2, is proportional to the ERR.
Fig. 9 indicates that the incidence of rockbursts
is also proportional to the ERR. The similarity
of Figures 8 and 9 demonstrates the value of using
seismicity as an indication of possible rockburst
incidence.

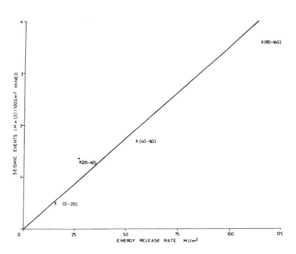

Fig. 8 Relationship between the number of
seismic events with M>1,5 per 1000 m^2 mined and
ERR as found at Blyvooruitzicht Gold Mine.

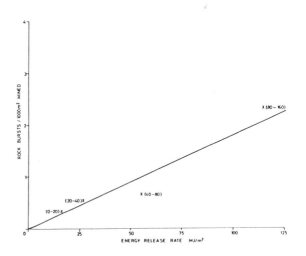

Fig. 9 Relationship between the number of
reported rockbursts per 1000 m^2 mined and ERR
as found at Blyvooruitzicht Gold Mine.

The linear relationship between the rate of
seismicity as defined above and the ERR implies
that the overall rockburst hazard encountered
while mining a given block of ground is governed
by the average ERR and does not depend on mining
sequence. Favourable initial mining layouts will
defer but not eliminate the rockburst problem.
This observation is of crucial importance as it
indicates that the key to the solution of the
overall rockburst problem is to control the
average ERR. In the case of Blyvooruitzicht
gold mine, as was mentioned earlier, this has
been done by implementing a system of stabilizing
pillars.

An important observation made at Blyvooruitzicht gold mine is that the rockburst losses, which are defined as the number of panel days' delay per 1000 m^2 mined, are relatively less severe in areas where a high ERR occurs (Fig. 10). Local ERR values of 80 MJ/m^2 to 160 MJ/m^2 have been encountered during the extraction of remnants, employing a method of up-dip mining. The explanation for this observation is that in remnant mining only one panel is worked and the face length is limited by the size of the remnant.

The effect of dykes and faults on seismicity near part of Blyvooruitzicht is illustrated in Fig. 11, which is a contour plot of the number of seismic events with M>1,5 per unit area, divided by the ERR to compensate for variations in mining geometry. Smoothing functions with dimensions of 100 m were applied to the data to reduce various spatial biases and errors. If the linear dependence of seismicity on the ERR (Fig. 9) were the dominant factor in controlling seismicity, there would be only minor variations in seismicity as shown in Fig. 11.

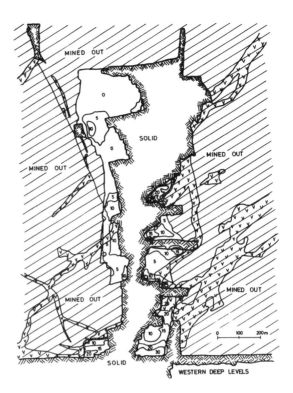

Fig. 11 ERR normalized seismic concentrations highlight the important influence of geological discontinuities on mining induced seismicity at Blyvooruitzicht gold mine.

Extensive geological features, as mapped routinely in the reef horizon, are also shown in Fig. 11. In regions of intense geological disturbances, the level of seismic activity is several times higher than in regions relatively unaffected by geological disturbances, Geological factors therefore appear to be even more important than geometric factors in controlling the rate of seismicity at Blyvooruitzicht.

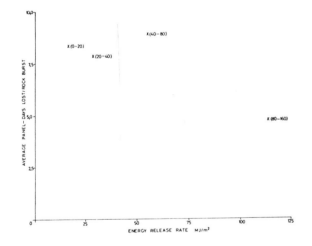

Fig. 10 Effect of ERR on the rockburst losses at Blyvooruitzicht Gold Mine

VAAL REEFS GOLD MINE

Vaal Reefs Exploration and Mining Company Limited is situated in the Klerksdorp goldfield some 160 km west of Johannesburg. The mining complex, which is the largest in the South **African** gold mining industry, comprises three **different** mining sections and produces approx**imately** 750 000 tons of ore per month.

The Klerksdorp area has proved to **be** very disturbed geologically (Fig. 12). Faults with throws of 10 m to 30 m are not extraordinary phenomena while throws in excess of 1500 m **also** exist in the area. As a result of this **faulting** the depth of mining varies from a few hundred

metres to more than 3000 m. The greatest depth
of stoping at Vaal Reefs is 2250 m.

Fault losses are excessive when compared
with the goldfields in the Far West Rand. The
actual percentage of ground loss is difficult to
ascertain, but it is believed to be between
15 per cent and 20 per cent of the total area.

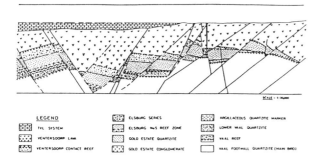

Fig. 12 Vertical section showing the geological
structure of the Klerksdorp district.

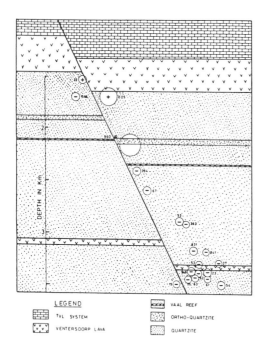

Fig. 13 In the Klerksdorp district it was found
that seismic events associated with fault planes
tend to occur on positions where the lithology on
either side of the fault plane is significantly
different.

Due to the large displacement on the faults,
coupled with the extent of the fault losses, it
is virtually impossible to undertake the longwall

system of mining in the Klerksdorp area and it
has therefore been necessary to resort to a
scattered mining system which is suitable to
extracting areas of reef between the faults.

Because of the faulted nature of the area,
the amount of off-reef development that has to
take place to prepare an area for stoping is
considerably greater than in other mining
districts. This, together with the fact that all
tunnels and haulages are developed prior to
stoping, makes the off-reef development very
vulnerable. A further factor which contributes
to the stability problem encountered with off-
reef development is the variable nature of the
footwall strata and the fact that because of the
faulted nature of the area most tunnels traverse
a variety of different strata.

Early in 1970 it was realized that the
pattern of seismicity in the Klerksdorp district
differed significantly from that observed on the
Central Witwatersrand. For this reason a regional seismic network which covered all four mines
in the district was established. With the aid of
this network it was possible to identify the
following typical event categories.

(i) Reef pillar events

These events occur in areas where unpayable
blocks of reef were left in situ. An
analysis of these events has shown that the
upper magnitude limit is a function of the
pillar area. Calculation of this is, however, difficult because of the complex pillar
geometry and the age of the pillar. The
upper magnitude limits for these events are
between 4,2 and 4,5 on the Richter scale.

(ii) Stope events

These events occur close to the working
areas and are generally in the lower magnitude range. The frequency of occurrence
increases during the middle of the weerk and
there is a marked decrease over weekends.
Another distinct feature of these events is
that they are closely related to the mining
induced energy changes.

(iii) Dyke events

Based on a detailed study of seismicity in
the Klerksdorp goldfield over a period of
nearly ten years, van der Heever[9] found that

approximately 50 per cent of a total of 2300 recorded seismic events were located within 100 m, 41 per cent within 60 m, and 30 per cent within 30 m of the nearest dyke. In general the events associated with dykes are smaller in magnitude than the reef pillar or fault type events. Dyke events with a Richter magnitude in excess of 4,0 are comparatively rare. It has been found that the more competent dykes tend towards greater seismic activity, while larger events are usually associated with the bigger dykes.

Although most dyke events are concurrent with active mining, the majority of the larger events occur on dykes with little or no seismic history, in areas remote from present mining activities.

(iv) Fault events

The largest damaging seismic events tend to occur near faults. Van der Heever[9] observed that there tends to be a close relationship between the dip component of the fault displacement, D, and the maximum probable magnitude, M_{max}, of seismic events:

$$\log D \text{ (in metres)} = 0,66 \, M_{max} - 0,85.$$

A particularly noticeable feature of seismic events which locate near fault planes is that they usually occur along these planes at positions where the lithology on opposite sides is significantly different (Fig. 13).

A study of the source mechanism of seismic events in the Klerksdorp district showed that events below a magnitude of 2,5 tend to have a higher proportion of first motion patterns with no clear fault plane solution. However, the majority of large events yielded reliable fault plane solutions which have focal mechanisms similar to those which occur during shallow crustal earthquakes where normal and reverse faults are reactivated. Underground observations of fault planes which are situated in the focal region of mine tremors revealed that renewed movement takes place across existing faults. The direction of this movement, however, is usually not the same as the sense of movement which occurred during the original faulting process.[9]

The damage associated with fault events is usually severe along the intersection of faults and service excavations and along the stope gullies which are nearest to the fault plane. Damage in the stopes is generally less severe, but because of the size of fault events scattered rockfalls can occur over large areas. Fig. 14 shows a typical example of damage sustained by footwall tunnel. Although the tunnel was badly damaged, the support fulfilled its function and access through this excavation was not lost. Figs. 15 and 16 show the unpredictability of damage done to on-reef excavations by an event with a magnitude of 4,6 which occurred on a fault with a displacement of 150 m. The two excavations are adjacent to the fault.

A feature of seismic damage in the Klerksdorp district is that it tends to affect off-reef excavations more severely than stoping excavations. Damage of the latter is often caused by widespread scattered rockfalls while complete stope closure as observed on some of the mines of the Central and Far West Rand is rare.

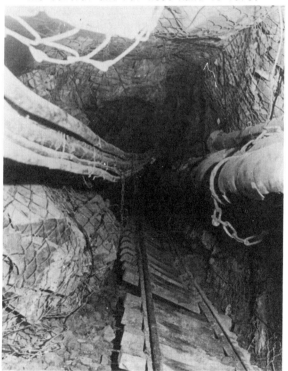

Fig. 14 Typical rockburst damage in a footwall tunnel in the Klerksdorp district

Fig. 15 Rockburst damage in a strike gully

Fig. 16 The absence of seismic damage in the stope panel next to that shown in Fig. 15 indicates the unpredictability of the area that will be damaged by an event.

MANAGEMENT STRATEGIES

The long-term objective of management strategy is to provide a mine layout that will enable the mine to be operated safely and profitably throughout its life. In order to achieve this, sound rock mechanics engineering is essential in the long-term planning of the mine, in the overall follow-up and in making short-term decisions which will be required where deviation from the mine standard system is necessary because of geological features or other disruptive factors. In the short term the main objectives of management strategy are to operate and control the mine safely and economically and to ensure that maximum planned mill throughput is achieved and maintained.

Rock Mechanics Department

Because of the important role rock mechanics engineering plays in the design and operation of deep level mines, most of the major gold mines have established rock mechanics departments. These departments assist mine management with long-term mine planning and mine design particularly as far as shafts and shaft pillars, service excavations, stoping layouts and the selection of support methods are concerned. In the short term the rock mechanics engineer and his staff assist with mine support and detailed planning of the extraction of ore from remnant and geologically disturbed areas.

The size and organizational structure of individual rock mechanics departments differ from mine to mine. In the case of Blyvooruitzicht gold mine the rock mechanics department is headed by a rock mechanics engineer. The rock mechanics engineer is responsible for the management and organization of the mine wide seismic network, the underground installation of haulage support and the hydraulic prop department. He is responsible for all standard instructions with regard to strata control, subject to the approval of the General Manager, and is actively involved in all planning aspects on the mine.

It was decided that the hydraulic prop department should report to the rock mechanics engineer for the following reasons:

- where props are dispersed over work sections control is necessary,

- the full and undivided attention of one or more officials is necessary,

- personnel with the necessary expertise are required,

- an autonomous organization ensures uniform prop standards.

At Vaal Reefs gold mine the rock mechanics department consists of a head of department, four rock mechanics officers and seven mine overseers with responsibilities for all the shaft systems. The latter are responsible for the scheduling of the off-reef support work and the quality control of the support of 8 km to 12 km of tunnels a month. The rock mechanics department at Vaal Reefs gold mine has direct access to the regional seismic office, operated by the Chamber of Mines Research Organization, which is situated at a neighbouring mine.

A total of 30 geophones cover the entire Klerksdorp gold mining district and enable locations to be made of seismic events with an accuracy of approximately 30 m to 50 m on plan and approximately 100 m in depth. Bimonthly meetings of rock mechanics and mining personnel from all four mines in the district ensure that changes in regional seismicity are recognized at an early stage and adequate precautions taken.

The fact that the rock mechanics department at Blyvooruitzicht gold mine emphasises stope support while the rock mechanics department at Vaal Reefs gold mine emphasises the support of tunnels and service excavations is indicative of the different nature of the rockburst problems on these mines.

Design of stoping excavations

Significant differences exist as far as the design of stoping excavations on the two mines is concerned. At Blyvooruitzicht gold mine the main emphasis is on the control of the ERR. As indicated on Figures 8 and 9, this is a good measure of the seismic problem. The entire mine is on a computer file and ERR figures are updated quarterly. These are used to determine critical areas and to plan stope support requirements. Because of the advanced stage in the life of the mine, stabilizing pillars rather than backfill have been employed to control ERR values. As far

as possible ERR values below 40 MJ/m^2 are aimed for. Whenever feasible dykes are incorporated into the regional support pillar system. This has two advantages. Firstly, mining of the dykes which is always dangerous and costly is greatly reduced and secondly, the gold losses in the stability pillars are minimized. An important and often neglected aspect of stability pillars is that they not only reduce the ERR but also limit the face length which is affected by large seismic events, thus eliminating one of the major drawbacks of the longwall mining system.

Fig. 17 shows a typical layout of a stoping section between strike stabilizing pillars. Six to seven 22,5 m long stope panels protected by 45 m wide pillars are advanced on strike. The stoping section is serviced by a so-called hangingwall drive which is situated below the upper strike pillar and two footwall drives at the bottom of the section. Footwall drive 'A' is required for the extraction of the bottom stope panel but becomes unserviceable about 100 m back from the face when it comes under the influence of the highly stressed strike pillar. Footwall drive B acts as a replacement haulage and is situated outside the influence zone of the strike pillar. In the reef plane itself a system of strike and dip gullies provides access to the working face.

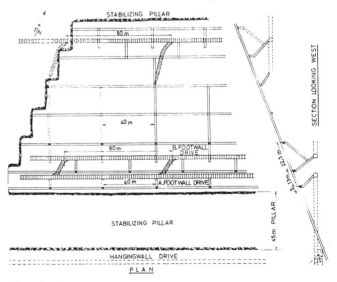

Fig. 17 Strike pillar layout and associated on reef and footwall development as implemented at Blyvooruitzicht Gold Mine.

This system of stabilizing pillars has been employed to good effect since 1979 and has made it possible to extract large remnant blocks between approaching longwalls in comparative safety (Fig. 7).

At Vaal Reefs gold mine the design of the stoping excavations is governed by the presence of many faults and dykes which divide the whole mining area into relatively small stoping compartments. As pointed out earlier, between 15 per cent and 20 per cent of the lease area is covered by fault losses. Because of this, mining spans are generally small and the ERR is consequently low. According to Table 2 almost 90 per cent of all stoping in the district has ERR values below 40 MJ/m^2. For the above reasons no regional stability measures either in the form of stability pillars or backfill are considered at this stage. Whenever possible stoping layouts are designed to minimize remnants between individual stoping connections.

By far the most serious problem at Vaal Reefs is stoping in the vicinity of faults. In the past, fault losses were considered as stabilizing or solid areas, with the result that in the scattered mining environment, mining was conducted towards the fault losses. This proved to be a mistake and resulted in seismically active areas where severe damage occurred in the on-reef excavations.

It was then decided that these situations could be avoided by mining away from the fault losses. With this method, it is necessary to have the access ways placed under the abutment formed by the fault, and on occasion traversing the fault itself. However, as the span of the stope increases, seismicity occurs and the access ways adjacent to the fault are usually lost.

As the seismicity associated with faults was not reduced when mining away from the faults, another method of mining adjacent to faults was investigated. The majority of the larger discontinuities in the Klerksdorp area lie parallel to the strike, thus presenting no major problem to mining along the discontinuity. By mining on strike, with the panels adjacent to the fault mined first and the remaining panels following in an underhand or overhand fashion, depending on the dip, the area adjacent to the fault is mined

at the lowest possible ERR.

As most of the seismic events associated with faults can be related to movement on the fault plane itself, experiments were conducted with different reef ribs against the fault. It was found that a rib which is five times the stoping width of an area is generally sufficient to stabilize a fault.

Extraction of remnant areas

A particular problem on any deep gold mine is the extraction of remnant areas. A number of routines have been developed to deal with this problem. These fall into three distinct categories, namely routines concerned with regional aspects of remnant extraction, local aspects and management aspects. In cases where more than one remnant is to be extracted the correct mining sequence is of utmost importance as the extraction of one remnant will have an influence on the stresses acting on other remnants. It is now standard practice to simulate the extraction sequence with one of the many computer programs that have been developed for the analysis of stress and energy changes induced by mining tabular ore deposits.

In order to concentrate mining activities, remnants are divided into groups. These groups are then mined in order of the magnitude of the stresses acting on them. Once the order of extraction has been finalized, the mining of each group is derived from this. The ERR of each remnant is calculated and the sequence that ensures the most uniform ERR for all remnants in the group is adopted.

A similar procedure is adopted when planning the extraction sequence of an isolated remnant. In general, the planned extraction sequence should be such that the formation of island remnants is avoided as far as possible; if this is not feasible, they should abut against solid blocks of ground. Wherever practical the remnant should be attacked from the narrow side. This ensures that the minimum length of face is exposed during the extraction period. This facilitates more uniform face advance and better support and limits the number of persons working under remnant conditions.

The following principles should be adopted when extracting remnants, in order to ensure safe working conditions.

The stoping width should be as narrow as possible to assist in lowering the ERR but at the same time should be such that even after sudden closure sufficient space will remain to conduct rescue operations.

All the faces should be blasted daily in order to extract the remnant as quickly as possible.

Advance per blast should not exceed 0,7 times the stoping width. This is necessary to enable the stress fractures ahead of the face to be established before the next blast and also assists in reducing the damage resulting from blasting.

Sequential blasting of shot holes should be ensured to maintain a straight face shape. This prevents stress concentrations on the face.

Active support, preferably rapid yielding hydraulic props, should be used as temporary support installed as close to the face as possible.

Rapid load bearing permanent support should also be installed as close to the face as possible, to ensure that all closure in the back area is used to generate load on the support.

Use should be made of well-designed regional support such as regional stabilizing pillars or waste fill.

Where possible, dykes should be left in situ unless they are likely to cause damage to ancillary excavations. If it is not necessary to stope a burst prone dyke and the induced stress along the edges of the dyke exceeds the uniaxial compressive strength of the material, a bracket or skin of reef may be left along the contacts to reduce the possibility of failure.

The direction of stoping should:

- be towards the largest solid area,
- avoid faults, dykes and other geologically disturbed areas, in adverse positions;

- make full use of accessible stope faces, cleaning facilities and travelling ways.

In addition sufficient access ways should be provided, and labour in the working areas must be kept to a minimum due to the prevailing dangerous working conditions. Additional support must be installed where faults and dykes are encountered, and if mining is to be towards the discontinuity, it must be done at a steep angle.

All these measures should be undertaken to ensure total extraction of a remnant, under safe working conditions. However, with the experience now gained, modern planning is aimed at preventing the formation of remnants and thus eliminating this hazardous mining situation.

Management policy as regards the mining of remnants has been found to be successfully implemented by means of Special Remnant Instructions. These instructions encompass the mining strategy outlined above in some detail. Remnants are defined so that the special remnant officer, who is appointed in writing by the General Manager, can easily identify each remnant and notify the relevant personnel.

The mining strategy in these defined areas is reviewed on a quarterly basis by a committee in order to co-ordinate mine policy. The decisions of this committee are communicated to all line management by means of plans and written instructions. This procedure ensures that management's expectations are communicated to all line management to ensure uniform standards in the mining of remnants.

Support

One of the most remarkable developments in deep level gold mining in recent years was the evolution of new support systems for gold mine stopes and off-reef excavations. All of these support systems are early load bearing, that is they provide support to the fractured strata around deep level excavations immediately or soon after installation, and at the same time have good slow and rapid yielding characteristics to accommodate the unavoidable deformations of deep mining excavations. Rapid yielding hydraulic props with a yield load of about 400 KN which are capable of yielding at a rate of more than 1 m/sec have been developed in the late 1960's to provide active support at the working face,[8].

Practical experience and recent theoretical studies have shown that the prop density is an important parameter as far as the control of stope damage due to seismic events is concerned. Under severe rockburst conditions prop densities in excess of 2,5 props per linear metre of stope face are required to confine stope damage to the immediate vicinity of the source of the seismic event[10]. In the back area of gold mining stopes the traditional timber mat pack was replaced by the timber and concrete pack which much improved early load bearing characteristics. More recently the so-called pipe stick, which is a 150 mm to 200 mm diameter wooden prop encased in a 3 mm to 4 mm thick steel pipe, has been introduced with very good results as permanent support in gold mine stopes. The excellent early load bearing characteristics, good yield properties and the ease of handling are the main features of this type of stope support.

The permanent stope support standards at Blyvooruitzicht followed the normal evolution from skeleton packs to solid timber packs to sandwich packs and finally to pipestick and rapid yielding hydraulic props.

Rapid yielding props were introduced on a limited scale in 1973, but when management decided in 1980 to introduce pipesticks on a mine wide basis, more than 20 000 rapid yielding hydraulic props were purchased to install three rows of rapid yielding hydraulic props in every panel on the mine.

The stope support system at Blyvooruitzicht was designed on the basis of active face support in conjunction with secondary internal passive support. The system is designed for an average face advance of 0,75 m per blast on a two blast cycle. Long axis gully packs were introduced at the same time as pipesticks to provide additional support to the gully ledges. The stope support system at Blyvooruitzicht has three main features:

the pipesticks generate a re-active load before the rapid yielding hydraulic props are moved forward,

pipesticks are installed under the protection of rapid yielding hydraulic props,

pipesticks can be installed closer to the face and are not affected by the blast because of the protection of the rapid yielding hydraulic props and the blast barricade.

The high frequency of faulting in the Klerksdorp district has given rise to a number of stope support problems. In the case of minor faults and şlips additional support is installed in either side of the discontinuity. This support must have rapid load bearing characteristics to prevent opening of the bedding planes particularly on the weak side of the discontinuity.

When mining in the vicinity of larger faults, associated faulting can cause severe support problems due to blocky ground conditions and high tectonic stresses. Experience has shown that faulting can add as much as 30 MJ/m^2 to the ERR that has been calculated for an area ignoring the effects of faulting. To compensate for the additional problems a good quality, stiff support must be installed. In certain areas at Vaal Reefs it is necessary to use rapid yielding hydraulic props as seismicity related to the faults can be extensive.

The development of support systems for gold mine tunnels has been less spectacular but equally effective. The passive timber or steel set tunnel support systems have been replaced by active rockbolt and grouted rock tendon support systems. The combination of these support elements with wire mesh and lacing ropes between the support tendons has resulted in an integrated support system which is capable of controlling most of the damage caused by even very large rockburst. The design of the actual support system depends greatly on local circumstances. For example at Vaal Reefs No. 5 shaft, which is situated in a large remnant formed by Hartebeesfontein gold mine, Buffelsfontein gold mine and Vaal Reefs itself, most of the development ends have to be supported by using 6 m grouted anchors. In addition to these anchors, which are installed on a 2 m diamond pattern, diamond mesh is placed against the rock surface and lacing cables are then installed through the protruding loops of the anchors. The effectiveness of this support type is clearly demonstrated in Fig. 14.

In some areas it has been found that 'grouted loops' or grouted ripple bars tend to snap under rockburst conditions. Smooth bar, grouted in, tends to be more effective in these areas. The friction between the cement and the smooth bar is not as high as the friction between the 'loops' or the ripple bar and the grout.

Blocky ground conditions often exist, creating problems with respect to development and support. The most effective support method for controlling these conditions is one which covers a large area, such as shotcrete or gunnite. Standard support for tunnels usually consists of:

rockbolts, used as temporary support and tensioned to 50 KN;

grouted loops, 2,3 m in length, installed on a 2 m diamond pattern with diamond mesh and lacing. These are installed to ensure stability under stress changing situations or moderate maximum principal stresses;

grouted loops, 3,2 m in length, installed with diamond mesh and lacing. These anchors cannot be installed in excavations smaller than 3 m x 3 m, therefore installation is restricted;

grouted loops, 6 m in length, installed with mesh and lacing. These are usually installed in areas already supported with 2,3 m grouted loops but where the stability of the excavation is endangered due to stress abutments advancing over the tunnel.

It is important for these support mediums to be installed before induced stresses endanger the stability of the excavation, as the effectiveness is greatly reduced if deterioration of the tunnel has already commenced.

CONCLUDING REMARKS
In this contribution it has been attempted to show that the magnitude and the nature of the rockburst problem on South African gold mines vary from district to district and indeed from mine to mine. It is for these reasons that no single, simple solution has been found for this pressing problem. The emergence of rock mechanics as an engineering discipline and profession together with the development of computer techniques to model complex tabular mining situations and the introduction of seismic technology into the industry provide an infrastructure on which mine management can base decisions on the method of mining, the risks involved and the eventual success of any mining operation. The development of this integrated approach to the rockburst problem is probably the most important step towards a solution of the rockburst problem that has taken place in recent years.

ACKNOWLEDGEMENTS
The contribution of the rock mechanics department at Blyvooruitzicht and Vaal Reefs gold mines to this paper and the permission of Vaal Reefs Exploration and Mining Company Limited to publish the paper are gratefully acknowledged.

References
1.   Analysis of working results (gold mining members of Chamber of Mines of South Africa). Published quarterly by the Chamber of Mines of South Africa, 1970-1980.
2.   An industry guide to the amelioration of the hazard of rockbursts and rockfalls. A Chamber of Mines of South Africa Publication. P.R.D. Series No. 216. 1977.
3.   Cook, N.G.W, Hoek, E, Pretorius, J.P.G, Ortlepp, W.D, and Salamon, M.D.G. Rock mechanics applied to the study of rockbursts. Journal South African Institute of Mining and Metallurgy, Vol. 66 1966, p. 435-713
4.   Hodgson, K, and Joughin, N.C. The relationship between energy release rate, damage and seismicity in deep mines. In: Proceedings of the 8th Symposium on rock mechanics, University Minnesota, A.I.M.E., New York, 1967, p. 194-209.
5.   McDonald, A.J. Seismicity of the Witwatersrand Basin, M.Sc. Thesis, University of the Witwatersrand, 1982.
6.   Salamon, M.D.G. Rock Mechanics of underground excavations. In: Advances in rock mechanics: Proceedings of the 3rd Congress, International Society for Rock Mechanics, Denver, Colorado. 1974, Vol. 1, Part B, p. 951-1099, Washington National Academy of Science, 1974.

7.   Salamon, M.D.G, and Wagner, H.   Role of
stabilizing pillars in the alleviation of rockburst hazard in deep mines.   In: Proceedings
of the 4th Congress,   International Society for
Rock Mechanics, Montreux, Switzerland, 1979
p. 561-566.

8.   Tyser, J.A. and Wagner, H.   A review of
six years of operation with the extended use
of rapid-yielding hydraulic props at the
East Rand Proprietary Mines, Limited, and
experiences gained throughout the industry.   In:
Papers and Discussions, Association of Mine
Managers of South Africa, 1976-1977, p. 321-348.

9.   van der Heever, P.K.   The influence of
geological structure on seismicity and rockbursts
in the Klerksdorp goldfield.   M.Sc. Thesis,
Rand Afrikaans University, 1982

10. Wagner, H.   Support requirements for rockburst conditions.   In: Proceedings of International Symposium on Seismicity in Mines.
Johannesburg. 1982.   To be published by
South African Institute of Mining and Metallurgy.

Fracture of rock at stope faces in South African gold mines

N.C. Joughin B.Sc. (Eng.), Ph.D., M.S.A.I.M.M.
Research Organization, Chamber of Mines of South Africa, Johannesburg, South Africa
A.J. Jager B.Sc (Hons.)
Geological Engineering Division, Mining Technology Laboratory, Research Organization, Chamber of Mines of South Africa, Johannesburg, South Africa

SYNOPSIS

A major advance in the control of rockbursts was achieved during the 1960's with the recognition that the rockmass around deep gold mining excavations behaves substantially as an elastic continuum. This enabled the incidence of rockbursts to be related to criteria such as stress levels and the energy released through potential energy changes in the rockmass due to mining. Subsequently, research has been concentrated on the mechanism of rockbursts and on the deviation of the rockmass from elastic behaviour because of fractures in the rock. It appears that rockbursts are of at least two distinct types, namely, those which are associated with movement on major geological faults and those which are associated with the formation of fractures at stope faces. This paper deals with the latter type and describes the progress made in understanding the formation of fractures in the rock at stope faces.

It has been established that the zone of fractured rock extends for a distance of many metres ahead of the face and that this distance is related to the energy release rate or stress level at the face. The fracture zone forms in direct response to the advance of the face and remains substantially unaltered when the face is left standing for a period of time. The fractures are formed stably without significant seismic activity. Disturbing seismic activity appears to be associated with geological irregularities in the rock which interrupt the normal stable formation of the fracture zone.

A detailed description of the types of fracture and of their behaviour within the fracture zone is given.

INTRODUCTION

In the deep gold mines of South Africa, rockbursts are a serious problem and have been researched intensively for many decades. A major advance in the control of rockbursts was achieved during the 1960's with the recognition that most of the rockmass affected by mining behaves substantially as an elastic continuum.[1,2] This enabled the stresses and displacements around excavations to be calculated and the rate at which energy was released per unit area of mining to be worked out for any depth and configuration of mining.[3,4] From this, and by seismic monitoring, it was shown that, in areas unaffected by complex geological structures, there were direct relationships between the energy release rate and the frequency and magnitude of seismic events.[5] From these findings and the recognition that there is a large number of seismic events that have hypocentres on, or close to, faults and dykes, or are concentrated at certain stratigraphic boundaries,[6,7] it was possible to introduce measures into the industry, during the 1970's, which have had the effect of either reducing the number of potential rockbursts or reducing the damage caused by any rockburst.[8,9]

Meanwhile, research into rockbursts continued actively. More than ten seismic networks with different purposes have been in operation at times.[10,11] Only work of direct relevance to this paper will be noted here. The main thrust of the relevant work was the understanding of the mechanism of rockbursts by examining seismic waves and by relating seismic activity to mining-induced fractures and pre-existing geological features.

Results of research into first motions of seismic waves emanating from rockbursts fall into two groups. In the one group, a preponderance of rarefactional first motions was reported suggesting the convergence of the surrounding rockmass on a volume of rock which failed suddenly near an excavation.[12,13,14,15] In the other group first motion radiation patterns which are consistent with a shear motion along a fracture plane were reported.[15,16,17,18] Such motion was attributed to slipping along pre-existing faults or dykes or the formation of "burst fractures". The interpretation of the two groups of results has been accompanied by some controversy. In a mining situation, first motion studies are difficult because there are relatively few recording stations which are often not favourably orientated towards, or ideally disposed about, the hypocentre of an event and which are often too close for a separation of the P and S waves. Nevertheless, the accumulated weight of evidence suggests that there are at least two types of mechanism, namely, the sudden failure of a volume of rock near the edge of an excavation and slipping along pre-existing faults or dykes which may be remote from an excavation.

P and S wave spectra have been used to investigate source dimensions and other parameters.[13,17,19,20]

Studies of fractures in the rock at the periphery of deep excavations have shown that there are several characteristic features enabling classification of the fractures.[21,22,23,24,25] However, in the different studies, there has been some inconsistency in the classification and interpretation of their mechanism of formation. One type of fracture, termed a "burst fracture", is presumed to be associated with violent seismic events in some of the studies. This type of fracture has an intensely pulverized surface a few centimetres thick, a shear displacement of a few centimetres, and extends for tens of metres. There has been some uncertainty about the association between "burst fractures" and seismic activity since, in some studies, there was no seismic network to confirm the association and, in others, the dimensions of the networks were such that the accuracy of location could not ensure an unambiguous association. Other problems have been that the rock stress conditions around the excavation were not taken into account

always, the method of roof support was not taken into consideration, and observations were made at the surface of the excavations without probing into the rock to determine the extent of the fractures. One notable exception was the tunnelling along a "burst fracture" to examine its extent.[26,27]

It became apparent that it was necessary to carry out a comprehensive study in which the formation of fractures could be observed continuously as mining progressed while seismic activity was recorded concurrently with greater precision and greater dynamic range. A fortunate opportunity presented itself for such a study. An investigation into the development of non-explosive mining methods was in progress at a site in the Doornfontein Gold Mining Company.[28] At this site it was possible to make detailed and repeated observations of mining-induced fractures without the complication of fractures caused by blasting. The face had a regular geometry, extending over a considerable length, which permitted study of fractures over a range of energy release rate conditions. Furthermore, this permitted drilling into the rock surrounding the face to probe the extent of fracture without undue interference from mining operations. Petroscopes were used for observing the formation of fractures, and extensometers were used for observing displacements in the fractured rock surrounding the stope faces.[29,30,31] Observations have been made in more than 300 drill-holes at Doornfontein and at other sites.

Three seismic networks of differing scale have been established at this research site. The largest network has dimensions of 2000 m x 2000 m and encompasses a large portion of the mine including the research site. It is capable of locating seismic events of magnitudes greater than $M_L = 0$ to an accuracy of 30 m. The intermediate network also encompasses the research site and has dimensions of 400 m x 400 m. It is capable of locating events of magnitudes greater than $M_L = -1,5$ to an accuracy of 10 m. The smallest network has dimensions of 16 m x 20 m and encompasses a short length of face within the intermediate network. It is capable of detecting events as small as $M_L = -5,5$ to an accuracy of about 3 m.

Figure 1. A photograph of the sidewall of a heading cut into the face of a stope, showing Type 1 fractures ahead of the face.

Figure 2. A photograph of the sidewall of a stope gully showing a Type 2 fracture in the footwall of the stope.

Observations have been in progress at the Doornfontein research site for a number of years now. A more distinct pattern of the behaviour of fractured rock and its association with seismic activity has emerged. Investigations have been extended to other sites to cover a wider range of circumstances. The results reported in this paper are essentially a summary of the more well established conclusions from all this work including the more important results that have been published. This paper is a companion to those presented at this meeting by Salamon on "The Rockburst hazard and the fight for its alleviation in South African gold mines" and by Wagner, Dempster and Tyser on "Regional aspects of mining induced seismicity - theoretical and management considerations".

CLASSIFICATION OF FRACTURES

The classification of fractures surrounding stopes is based principally on their inclinations, on the presence or absence of movement across them, and on secondary associated structures. Three types of fractures are recognized. All the fractures strike substantially parallel with the stope face.

Type 1 fractures

Type 1 fractures are by far the most common, Figure 1. They are planar, steeply inclined, and have no displacement in the plane of the fracture when first they are formed. The inclination varies with the energy release rate or span of the stope, but for large spans ($>$ 150 m) they are inclined at angles of between 80° to 100° to the horizontal. For smaller spans the fractures in the immediate hangingwall and footwall usually are inclined at shallower angles in the direction of face advance.[22] Depending on the energy release rate, these fractures form several metres ahead of the stope face in zones, each zone consisting of 5 to 20 fractures and separated from its neighbouring zone by a width of intact rock. In later stages of the development of the fracture zone, as the face approaches close to fractures formed earlier, further fracturing takes place to form parallel-sided slabs and small angled wedges of rock. Convergence of the hangingwall

55

and footwall causes relative movement between the slabs and concomitant horizontal displacement of fractured rock towards the stope. These movements cause comminution along the fracture surfaces and secondary minor cross fracturing from type 1 fractures to close-by adjacent fractures, leaving the rock in a fairly cataclastic condition.

At their time of formation, type 1 fractures are almost certainly aligned with the maximum principal stress direction. Conventional stress measurements in the region of high stress immediately beyond the fracture front ahead of a stope cannot be made because of the fracturing associated with the borehole and the discing of the core. Fractures which resemble type 1 fractures can be produced in laboratory extension test apparatus. These fractures tend to form when the minimum principal stress at failure is low to moderate.

Type 1 fractures occur in all stopes where the stress levels are sufficient to induce fracturing and their incidence increases as the energy release rate increases. They usually terminate at stratigraphic parting planes although many may extend through a number of parting planes.

Type 2 fractures

Type 2 fractures are far less common than type 1 fractures and are inclined at angles of 60° to 75° to the horizontal, Figure 2. They have a component of displacement up to 150 mm in the plane of the fracture. These fractures can be quite complex. In their simplest form, they are a few millimetres wide with displacements of a few millimetres. The fault zone is filled with highly comminuted material. Usually the fault zone is several centimetres wide. In the more complex structures, there are several major fractures en echelon where the region between overlapping portions of the main fractures is brecciated by cross fractures. Sometimes conjugate feather fractures are formed, resulting in a fractured band more than 500 mm wide. Large fractures have been traced for 30 m along their strike, and for several metres out of the plane of the reef.

In the laboratory, fractures can be produced in extension test apparatus which resemble simple type 2 fractures. These fractures are formed when the minimum principal stress at failure is much higher than that when fractures resembling type 1 are formed. The implication is that type 2 fractures are formed underground in circumstances where the confining forces are high. Supporting this is the observation that type 2 fractures do not occur at energy release rates below about 10 MJ/m^2.

Type 2 fractures appear to be similar to, if not identical with, the "burst fractures" described by other investigators. A very important conclusion from the seismic recordings at Doornfontein is that these fractures formed without noticeable seismic activity. It is therefore incorrect to assume that they occur violently.

Type 3 fractures

Type 3 fractures have an inclination of 20° to 40° to the horizontal towards the direction of face advance, Figure 3.

Figure 3. A photograph of a hangingwall exposure showing Type 3 fractures inclined at 20°.

They are clean planar fractures, which display no evidence of shear displacement, are confined to the first few strata above and below the stope, and seldom have strike lengths greater than 5 m. Type 3 fractures form close to the stope face within rock which had previously experienced Type 1 and Type 2 fracturing. The type 3 fractures however develop preferentially in the previously intact zones separating the fracture zones which formed several metres ahead of the face. They are thus a second generation of fractures superimposed onto the more regional fracture pattern. They appear to occur in response to the very localized stress changes at the periphery of the stope excavation resulting from the horizontal displacement of fractured rock from the fracture zone towards the mined out area.

Sometimes other types of fracture form such as fractures parallel to the bedding in the immediate hangingwall and footwall.

NATURE OF THE FRACTURE ZONE

The fracture zone around a stope face exhibits a remarkably systematic pattern of behaviour once the main factors influencing it have been recognized and taken into consideration. Figure 4 is a conceptual representation of the fracture zone around a stope face. The following is a description of the main features of the fracture zone.

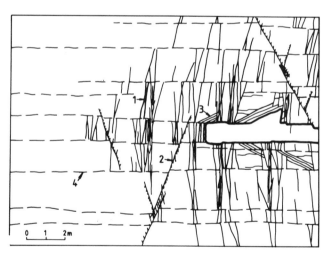

Figure 4. A diagram of the fracture zone around a stope face showing Type 1, Type 2, Type 3 fractures and parting planes (4).

Development of the fracture zone

Most of the fractures develop in direct response to the mining of the stope face. Thus, as the face is advanced, the fracture zone migrates forward ahead of the face. It appears that the new fractures form within a very short time after advancing the face and that very little fracturing takes place more than 3 hours after the face has been advanced.

The normal development of the fracture zone is stable without the emission of significant seismic energy. Most of the seismic energy is sub-audible and only when the stope is very quiet can clicks or cracks be heard. The events located on the smallest seismic network and which can be associated with the stable formation of the fracture zone radiate energy much less than a joule.[32] The normal development of the fracture zone cannot be detected on the intermediate and large networks.

The abnormal development of the fracture zone is discussed later.

Dimensions of the fracture zone

The distance to which the rock is fractured ahead of the stope face is dependent principally on the energy release rate, but may also be affected to a lesser extent by factors such as rock type and the stiffness of roof support. Thus, at an energy release rate of 10 MJ/m^2, fracturing will extend between 1 m and 2,5 m ahead of the face; at 20 MJ/m^2, between 2,5 m and 5 m ahead; and at 45 MJ/m^2, between 4,5 m and 7 m ahead of the face.

The distance to which fracturing occurs above and below a stope has not been established fully. Probably, this distance is also proportional to the energy release rate. At a site where the energy release rate was 25 MJ/m^2, stope-related fractures were observed down to a depth of 45 m below the stope. Fractures at this distance from the reef plane are not continuous with those occurring directly ahead of the stope but are rather confined to an individual stratum bounded by parting planes. Fractures therefore can initiate tens of metres above or below a stope.

The intensity of fracturing decreases with distance from the reef but it has been found that a particular stratum with a high propensity for fracture can occur in a state that is more fractured than that of a stronger stratum closer to the reef. Close to the stope elevation where the stresses are higher, fractures may propagate across a number of strata but few fractures have a vertical extent of more than 4 m.

Disposition of fractures in the fracture zone

The majority of fractures are substantially parallel to the stope faces. Where there are marked changes in the direction of a face such as at a corner abutment, the fractures follow the face around the corner except at the apex of the corner where the fractures follow a curved path suggestive of the maximum principal stress direction. In the same way, fractures do not follow minor irregularities in the face. Where there are regular deviations in the shape of the face, such as the step-like pattern between successive panels on a longwall, and where the energy release rate is sufficiently high, long pronounced type 1 fractures, which follow the overall face shape, form in addition to the many normal fractures which follow the individual panel faces closely.[25]

The type 1 fractures ahead of the face cluster in bands which are separated from each other by relatively unfractured rock. The width of the fractured bands is typically 500 to 700 mm and that of the intervening intact rock 400 to 800 mm. The length of the bands parallel to the stope face is of the order of 15 m but slightly displaced extensions of such zones can continue en echelon in the dip direction for over 40 m. The spacing between individual type 1 fractures ahead of the face appears to be independent of the energy release rate. However, it is thought that this spacing may be affected by other factors such as the stope height and the type of roof support used. At the Doornfontein site, where the mined out area is filled with barren rock, the median spacing of the type 1 fractures is 60 mm. While the inclinations of most fractures near the stope are close to vertical, there is a regular decrease in the dip of fractures with distance from the reef such that, at 45 m below the stope, the fractures are inclined at 55° in the direction of face advance. From limited evidence it appears that the fracture front is furthest ahead of the stope in the plane of the stope and runs backwards under and over

the stope at angles of between 50° and 60°. It is thought that, under normal conditions, the whole front advances with the advance of the face.

Displacement of fractured rock within the fracture zone

There is considerable relative motion, in a vertical sense, between fragments of rock in the zone ahead of the face as evidenced by faulting of boreholes drilled ahead of the face. The relative motion can take place along type 2 fractures or, in the later stages of the development of the fracture zone, along type 1 fractures. Displacements of more than 25 mm have been observed.

Pronounced movement takes place in the horizontal direction. All the fractured rock is forced away from the solid zone towards the mined out area. Movement commences slightly ahead of the fracture zone where the rock is approaching a state of failure. As the rock undergoes failure, it dilates laterally. This dilatation, equivalent to 5 to 10 millistrains, amounts to a horizontal displacement of a few millimetres.

Nearer the face, where the rock is in the later stages of fracture development, gross dilatation of about 10 to 80 millistrains takes place. In this region the dilatation is increased because of the interleaving of type 1 fragments and the lateral motion resulting from movement along type 2 fractures.

The rock between the two regions of dilatation undergoes little fracture and is merely translated bodily towards the excavation. However, since the extents of fracture in adjacent strata may differ, the horizontal movement in adjacent strata may differ resulting in considerable slip at the parting planes separating the strata.

Horizontal movement towards the mined out area also takes place through the rotation of slabs of fractured rock, created by type 1 fractures, in the immediate hangingwall and footwall. Such motion is particularly pronounced when gullies are cut parallel to the face in the hangingwall or footwall.

It is obvious that the roof support used in the stope has an effect on the movement of the fractured rock towards the mined out area. It is equally obvious that this could have a major bearing on the development of fractures as it would alter the confining forces on the rock undergoing failure. Investigations into this aspect are still in progress and it is premature to draw conclusions about the effects of support on the fracture zone.

Influence of geological structures on the fracture zone

Joints, faults, and dykes have marked effects on the development of the fracture zone. The orientation of any of these structures with respect to the stope face is important. Their effects are least when they are at right angles to the face, and greatest when parallel to it.

In the area of closely spaced ($<$ 400 mm), parallel, open jointing it has been observed that there has been an almost complete absence of mining-induced fracturing. In dykes, which often are jointed in more than one direction, the intensities of mining-induced fracturing may be significantly less than in the adjacent quartzites. It is not uncommon to find that the density of fracturing on one side of a fault is very different from that on the other side. The explanation for this may be in anomalous tectonic stresses in the vicinity of the fault[33] or a distortion of the induced stress by a particular orientation of fault relative to the face direction.

These interruptions in the regular development of the stable fractures appear to be significant to the generation of unstable failure in that they allow the build up of abnormally high stresses. This will be discussed in more detail in the section on the mechanism of seismic events.

SEISMIC ACTIVITY

Location of seismic activity

The large seismic network at Doornfontein has revealed a pattern of seismic activity no different from that on mines with similar geological structures. Most seismic activity occurs close to the plane of the reef and is concentrated on stope faces in areas having high energy release rates and near dykes and faults, Figure 5. The highest concentrations are in remnant areas

intersected by dykes. The largest event located on this network had a magnitude of M_L = 3,3.

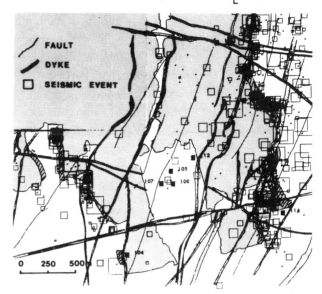

Figure 5. A plan showing the positions of seismic events located by the large seismic network at the Doornfontein gold mine during one year. The size of each square represents the magnitude of the relevant event.

The intermediate seismic network has been most revealing in associating seismic activity with specific geological structures. In the area encompassed by this network, three geological structures are of significance.[34]

a) The most prominent structure is a dyke, 3 m to 8 m wide. Joints and faults with displacements of up to 2 m are associated with this dyke which is almost vertical and intersects the stope face at an angle of about 60°.

b) A system of parallel quartz veins occurs throughout the area. They are spaced at intervals of less than 5 m and intersect the stope face at an angle of about 20°. The quartz veins, although a prominent feature, are strongly bonded with the quartzite and therefore do not represent pronounced planes of weakness in the rock.

c) There is a group of open joints and normal faults which intersect the face at an angle of -40° to -50°. The two most prominent faults in the group have displacements of less than 200 mm. They are 60 m apart and are inclined at 50° to the reef plane. The joints are distributed erratically between the faults. From a mining point of view, these faults are insignificant and would not be recorded during routine geological mapping.

Figures 6 and 7 show the pattern of seismic activity located by the intermediate network. The smallest events located with this network were of magnitude $M_L = -1,5$; the largest event located had a magnitude of $M_L = 2,8$ while the commonest events had magnitudes of $M_L = -1$. The most striking feature is that the seismic activity occurs close to the stope face and follows the group of minor faults and joints. There seems to be no association with either the quartz veins or dyke. Although it is not apparent fully in Figure 5, there was increased seismic activity and considerable difficulty was experienced when the face at the intersection of the dyke and the group of joints was mined.

fracture observed which displayed characteristics different from those of the three types described. This was despite the occurrence of events with magnitudes of up to $M_L = 2,8$ being located close to the stope face. All three types of fracture occurred throughout the area in similar relative proportions.

The only obvious difference in the fracture pattern between the seismically active and inactive areas was that where the joint spacing was less than 400 mm, the mining-induced fractures were significantly fewer and in some places totally absent. This implies that the joints inhibited the normal development of the fracture zone and that the seismic activity might have been the sudden re-establishment of the fracture zone.

Since it is well known that severe rockbursts are associated with dykes, particularly where stope faces are almost parallel with them, the only conclusion that can be drawn here is that as the dyke in the research site was inclined at a large angle to the face, it caused little interference with the development of the fracture zone. However, this dyke did give rise

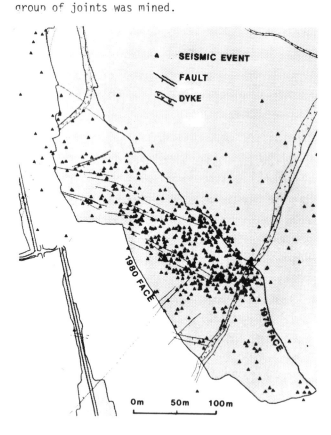

Figure 6. A plan showing the positions of seismic events located by the intermediate seismic network at the research site over a five year period.

At no stage during the mining of the area encompassed by the intermediate network was a

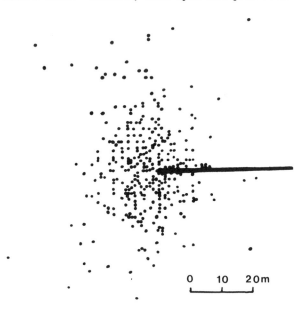

Figure 7. Distribution of seismic activity around the stope face; sectional elevation of a stope showing the hypocentres of seismic events in Figure 6 plotted relative to a common face position.

to additional problems where it intersected the group of joints which gave rise to the seismic activity.

Investigations using the smallest seismic network are still in the early stages and it is premature to draw any definite conclusions. However, it is of significance that patterns of activity have been observed which suggest the formation of type 2 fractures, Figure 8.

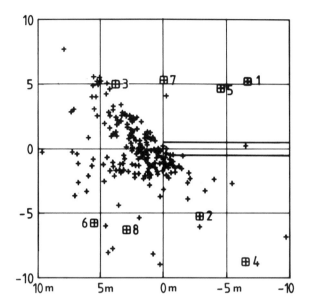

Figure 8. A section looking up dip showing the positions of seismic events located by the small network during one day.

Since type 2 fractures occurred throughout the area encompassed by the intermediate network, including the parts where no seismic activity greater than $M_L = -1,5$ was observed, it can be concluded that these fractures can form stably by the coalescence of innumerable small cracks.

While the results of the intermediate seismic network show a distinct association between geological structure and seismic activity, it could be erroneous to extrapolate these conclusions to the areas of concentrated seismic activity identified in the large network. In the area encompassed by the intermediate network, the energy release rate was moderate, ranging from 10 MJ/m^2 to 30 MJ/m^2, while the energy release rate was very much higher in the areas of concentrated activity in the large network. Furthermore, the geometries of the areas of concentrated

activity in the large network were very complicated; the remnants being of irregular shape and having fracture zones on all sides. Since it is apparent that minor geological structures can give rise to violent seismic activity, a seismic network with accuracy better than that of the intermediate network would be required to identify the factors which give rise to rockbursts in remnants.

EXAMINATION OF ROCKBURSTS

Of the 669 seismic events located within the intermediate network, nine caused rockbursts of varying severity. The smallest event to cause a rockburst had a magnitude of $M_L = 0,5$ and the largest $M_L = 2,8$. Most of these events were located ahead of the face and had hypocentres within 20 m of the face and the reef plane. It should be noted that the errors of location of the larger events were often greater than those of smaller events. All the rockbursts occurred within the seismically active region associated with the group of minor faults and joints.

The nature of the damage caused by these rockbursts was found to be similar for all cases although the length of face damaged ranged from 4 m to 75 m. The minor faults had a controlling influence on the extent of face damage for, in every rockburst, either one, or both, of the limits of the damage coincided with a fault. Blocks of rock up to 1/4 m^3 in size were ejected violently from the face for distances of up to 6 m. In most cases, rock, to a depth of less than 0,5 m, had been ejected from the face but, in some instances, as much as 1 m was ejected from the face. The rock left on the face after a burst always was fractured intensely.

The most important feature common to all rockbursts was marked convergence between the hangingwall and the footwall. The greatest convergence observed was 250 mm in the case of the largest rockburst. During the mining of the face subsequent to this rockburst, there was evidence that the rock ahead of the face had been crushed for a distance of 3 m. The crushing was greatest at the face and diminished gradually over the 3 m. This could be seen from the step-like displacements between type 1 fractures which intersected parting planes in the hangingwall and footwall and which were exposed during the subsequent mining. It was apparent that movement in the vertical direction had taken place along the type 1 fractures causing interleaving of the fractured rock which, in turn, forced the fractured rock horizontally into the excavation. In all the rockbursts, displacement along type 1 fractures was apparent, but, only in two of the rockbursts was there evidence that displacement had occurred along type 2 fractures. Also, for all the rockbursts there was no obvious movement on faults and joints except in the case of the largest rockburst. Here, where a small fault intersected the face near the centre of damage, convergence had resulted from the downward movement of the hangingwall on the one side of the fault and the upward movement of the footwall on the other side of the fault.

In all the rockbursts, damage coincided fairly closely with the area of convergence. Secondary damage in the form of falls of rock from the hangingwall was experienced in all cases. These falls, although a consequence of the accelerations imparted to loosened, fractured rock, usually have no bearing on the mechanism of the rockburst and often occur in stopes which have not suffered authentic rockburst damage. Type 3 fractures contribute greatly to secondary damage. In the case of the largest rockburst, extensive falls occurred as a result of an unusually long type 3 fracture which extended for the full length (75 m) of the damaged area.

The roof support had a marked effect in controlling the secondary damage. Most of the damage occurred near the face and seldom extended beyond the first line of rapid yielding hydraulic props and no damage extended into the area supported by backfill.

Mechanism of seismic events

Investigations into the mechanism of seismic events have been confined to the intermediate network on account of the simple geometry of the stope face and the very detailed observations regarding geological structures and fractures. First motions of P-waves have been analysed for 96 events which were located within 20 m of the face in the seismically active part of the face bounded by the group of minor faults and joints.[36]

Only first-motion data which were indisputably clear were used. Consequently no more than eight data points were obtained for each event. With such paucity of data, it would be possible to postulate more than one mechanism for each event. However, on comparing the data for all the events, it was apparent that there were certain similarities between the radiation patterns of the events. The first-motion data points for individual events were stacked by superimposing stereographic projections of the data points along the same selected axes. Different axes, chosen to have a physical relationship to the stope geometry, were used. For instance, one axis chosen was along the strike direction of the faults. Thus, if the mechanism of all the seismic events was slippage along the fault in the dip direction, then stacking of the data points should reveal a clearer, four-lobed radiation pattern.

Stacking of first-motion data points along axes in the fault plane yielded no consistent patterns. The plot turned out to be a thorough intermingling of compressions and rarefactions from which it could be concluded that the mechanism for the events was not one of slippage along the faults. Underground observations corroborated this conclusion as no evidence of recent movement on the faults could be found except in the case of the largest rockburst where the hangingwall had moved downwards on the one side of the fault and the footwall had moved upwards on the other side of the fault. Unfortunately first-motion data for this event were indistinct.

The only direction of the axis of projection which yielded a meaningfull pattern was parallel with the stope face. This direction has considerable physical significance since it is parallel with all three types of fracture. Projection along this axis showed similar patterns for about half the events. Stacking of the first-motion data for these events is shown in Figure 9. The radiation pattern that results is approximately two-lobed with rarefaction towards the mined area and compression towards the unmined area. Similar first-motion data observed

on an adjacent stope face within the area encompassed by the network, lend credence to this
radiation pattern. The adjacent face was being
advanced in a direction opposite to that of the
research site. The radiation pattern was similar,
but reversed, Figure 10, corresponding to the
advance of the face in the opposite direction.

pattern. This implies that no sudden slipping
along type 2 fractures occurred close to the face.
Nevertheless, it is acknowledged that there is
no certainty that some violent events are not due
to slippage on type 2 fractures or faults. With
the knowledge that the fracture zone can be very
large and that complex displacements can occur

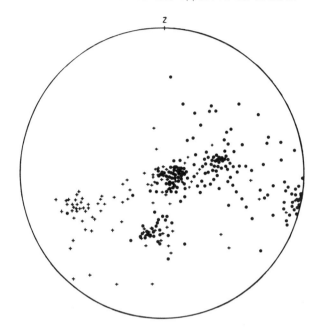

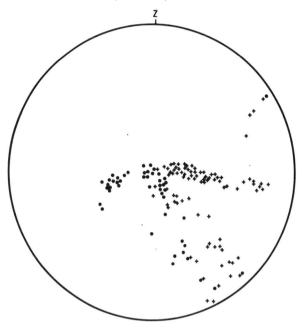

Figure 9. P-wave first-motion data in a
composite stereographic plot looking along an
axis parallel with the stope face. The stope is
to the right of the centre point. Dots represent
rarefactions and crosses represent compressions.

Figure 10. Similar to figure 9 except that the
stope is to the left of the centre point.

A mechanism which could give rise to this
type of radiation pattern is the sudden failure
of a volume of rock on one side of a boundary.[13]
Such a mechanism would be equivalent to the
sudden failure of a volume of rock ahead of a
stope and would be completely consistent with the
observations regarding convergence and crushing
of the rock ahead of the face during rockbursts
and the interruption of the normal development
of the fracture zone by geological irregularities.

Attempts were made to classify the remaining
half of the events into groups having similar
radiation patterns. This yielded some rather
unconvincing patterns which could be interpreted
as the sudden failure of blocks of unfractured
rock within the fracture zone. However, it is
more important to note that these events could
not be grouped to form a four-lobed radiation

within it, it is possible to visualize a host of
mechanisms involving slipping on multiple fractures which could yield a variety of radiation
patterns including the two-lobed pattern. Also,
no explanation of the mechanism has been offered
for the many events located far from the face
both inside and outside the fracture zone. These
may well be slipping on fractures and faults.

ENERGY DISSIPATION

It has long been recognized that a key factor in
understanding violent seismic activity is an
understanding of the mechanism of stable dissipation of energy released by mining. A widely held
view is that the incidence of rockbursts is proportional to the amount by which the energy release rate is greater than the rate at which
energy can be dissipated non-violently as the
excavation is enlarged.[7] The assumption here was
that only a limited amount of energy could be
dissipated non-violently and that the remainder,

of necessity, would be dissipated violently.

In all previous attempts to explain the stable dissipation of energy released by mining, the extent and behaviour of the fracture zone were not appreciated. This study has shown that violent seismic activity is related to the abnormal development of the fracture zone. This implies that, in the normal development of the fracture zone, the energy released can be dissipated in a stable manner over a wide range of energy release rates, and that the dimensions of the fracture zone increase to accommodate higher energy release rates. Analytical models which simulate the behaviour of the fracture zone are in the process of being developed.[37,38] With these models, taking into account the degree of fracture and the displacements which occur within the fracture zone, it can be shown that dissipation of all the energy can be accounted for by friction.

SUMMARY

Understanding of the behaviour of the zone of fractured rock surrounding stope faces has been increased considerably. For the most part, the formation of the fracture zone is a stable and systematic process. Violent seismic activity is abnormal and seems to be due to irregularities in the geological structure which interfere with the normal formation of the fracture zone.

The fractures within the fracture zone have distinctive features which make it possible to classify them into three main types. All three types are parallel to the stope face but they differ in their orientations, surface textures, and displacements along the fracture surfaces. All three types normally form stably without significant seismic activity, but since they occur in seismically active areas also, it is not certain whether they always occur stably. In the normal development of the fracture zone, new fractures form, within a short time, in response to advancing of the face.

Considerable movement occurs across fractures and bedding planes within the fracture zone. Factors, such as roof support, stope heights, and gullies, influence the mobility of the fractured rock. In turn this would influence the confining forces and the fracture processes within the fracture zone. The movements within the fracture zone are sufficient to account for the dissipation by friction of all the energy released by mining.

The overall dimensions of the fracture zone are closely related to the energy release rate. Where the energy release rate is very low, the fracture zone is small; where the energy release rate is 40 MJ/m^2 the fracture zone extends to more than 5 m ahead of the face.

At the Doornfontein mine, the pattern of seismic activity is no different from that on other deep gold mines. The activity is concentrated in areas where energy release rate is high because of irregular geometries, particularly areas intersected by dykes and faults. At the research site, with the long straight face and moderate energy release rate, the seismic activity was confined to a small part of the face intersected by a group of minor faults and joints. The only difference in behaviour of the fracture zone that was observed in the active part of the face was that the fractures did not form regularly when the face was advanced.

Investigations into the mechanism of seismic events revealed no evidence of slipping along faults. Instead they indicated the sudden failure of a volume of rock ahead of the stope face.

It is now well established that the occurrence of rockbursts and violent seismic events can be reduced by controlling the energy release rate. The conclusion reached here, that rockbursts and violent seismic activity are not the normal accompaniment of mining at depth, is important because it suggests that alternative measures for reducing their occurrence might exist by deliberate modification of the fracture zone. It is premature to foresee what these measures might be; nevertheless some possibilities suggest themselves. For instance it may be possible that, by adopting a particular form of roof support, the fracture zone might be modified in such a way that it is less sensitive to structural irregularities. Another possibility is that, simply by recognizing where the fracture zone is not developing normally, fracture may be induced by artificial means.

ACKNOWLEDGEMENT

The investigations described were conducted as part of the research programme of the Mining Technology Laboratory in the Research Organization of the Chamber of Mines of South Africa. The part played by the management and staff of Doornfontein Gold Mining Company, Limited, in supporting the investigations is acknowledged gratefully.

References

1. Ortlepp W.D. and Cook N.G.W. The measurement and analysis of the deformation around deep, hard rock excavations. Proceedings Fourth International Conference on Strata Control and Rock Mechanics, Col. Univ., New York, (1964) p. 140-150.

2. Ryder J.A. and Officer N.C. An elastic analysis of strata movement observed in the vicinity of inclined excavations. J.S. Afr. Inst. Min. Metall. vol. 64, (1964) p. 219-244

3. Salamon M.D.G. Elastic analysis of displacements and stresses induced by the mining of seam or reef deposits. Pt. I; II; III. J.S. Afr. Inst. Min. Metall., vol. 64 (1964) p. 128-149; 197-218; 468-500.

4. Plewman R.P. , Deist F.H. and Ortlepp W.D. The development and application of a digital computer method for the solution of strata control problems. J.S. Afr. Inst. Min. Metall., vol. 70 (1969) p. 214-235.

5. Hodgson K. and Joughin N.C. The relationship between energy release rate, damage and seismicity in deep mines. Proceedings Eighth Symp. on Rock Mechanics., University of Minnesota (1966).

6. Van der Heever P.K. A seismic investigation of mine tremors in the Klerksdorp mining complex. S. Afr. Assoc. Mine Managers Circ. No. 2/78(1978).

7. Cook N.G.W., Hoek E., Pretorius J.P.G., Ortlepp W.D. and Salamon M.D.G. Rock mechanics applied to the study of rockbursts. J.S. Afr. Inst. Min. Metall. 66, (1966) p. 435-528.

8. McGarr A.and Wiebols G.A. Influence of mine geometry and closure volume on seismicity in a deep-level mine. Int. J. Rock Mech. Min. Sc. and Geomech. Abstr., vol. 14, (1977) p.139-145.

9. Tyser J.A. and Wagner H. A review of six years of operation with the extended use of rapid yielding hydraulic props at the East Rand Proprietary Mines Ltd. and experience gained throughout the industry. Assoc. Mine Managers. S. Afr. Papers and Discussions 1976/1977 p.321-347.

10. Chamber of Mines of South Africa. Research and Development Annual Reports. 1975 to 1982.

11. The Proceedings. Rockbursts and Seismicity in Mines. Editors Gay N.C. and Wainwright E.J. Symp. Series No. 6. S. Afr. Inst. Min. Metall. Johannesburg (1983).

12. Gane P.G., Seligman P. and Stephen J.H. Focal depths of Witwatersrand Tremors. Bull. Seis. Abs. vol. 43, No. 3, (1952).

13. Joughin N.C. The measurement and analysis of earth motion resulting from underground rock failure. Ph.D. Thesis Univ. of Witwatersrand (1966)

14. Hallbauer D.K. Unpublished report. (1967).

15. Van der Heever P.K. The influence of geological structure on seismicity and rockbursts in the Klerksdorp gold field. M.Sc.Thesis Rand Afrikaans University 1982.

16. Spottiswoode S.M. Source mechanism studies on Witwatersrand seismic events. Ph.D. Thesis. University of the Witwatersrand (1980).

17. Spottiswoode S.M. Source mechanism of mine tremors at Blyvooruitzicht Gold Mine. Rockbursts and Seismicity in Mines. Editors Gay N.C. and Wainwright E.J. Symp. Series No.6 S. Afr. Inst. Min. Metall. Johannesburg (1983).

18. Potgieter G.J. and Roering C. The influence of geology on the mechanism of mining associated seismicity in the Klerksdorp gold fields. Rockbursts and Seismicity in Mines. Editors Gay N.C. and Wainwright E.J. Symp. Series No. 6 S. Afr. Inst. Min. Metall. Johannesburg (1983)

19. Spottiswoode S.M. and McGarr A. Source parameters of tremors in a deep-level gold mine. Bull. Seism. Soc. Am., vol. 65, (1975) p. 93-112.

20. McGarr A. Some applications of seismic source mechanisms to assessing underground hazard. Rockbursts and Seismicity in Mines Editors Gay N.C. and Wainwright E.J. Symp. Series No. 6. S. Afr. Inst. Min. Metall.

Johannesburg (1983).

21. Pretorius P.G.D. Some observations on rock pressure at depth on the E.R.P.M. Ltd. Ass. Mine Managers of S. Afr. (1958) p. 405-446

22. Kersten R.W.O. Structural analysis of fractures around underground excavations on a Witwatersrand gold mine. M.Sc. Thesis University of Pretoria. (1969).

23. McGarr A. Stable deformation of rock near deep-level tabular excavations. J. Geophys. Res. Vol. 76 No. 29, (1971) p. 7088-7106.

24. McGarr A. Violent deformation of rock near deep level tabular excavations. Bull. Seismol. Soc. Am., vol. 66, (1971) p. 685-700.

25. Hagan T.O. A photogrammetric study of mining-induced fracture phenomena and instability on a deep-level longwall stope face with variable lag lengths. M.Sc. Thesis Rand Afrikaans University (1980)

26. Ortlepp W.D. The mechanism of a rockburst. Proceedings 19th U.S. Rock Mechanics Symp., Reno NV. (1979) pp. 476-483.

27. Gay N.C. and Ortlepp W.D. The anatomy of a mining-induced fault zone. Bull. Geol. Soc. Am. vol. 90 (1979) pp. 47-58.

28. Joughin N.C. Progress in the development of mechanized stoping methods. J.S. Afr. Inst. Min. Metall. (1978) p. 207-217

29. Van Proctor R.J. An investigation of the nature and mechanism of rock fracture around longwall faces in a deep gold mine. Ph.D. Thesis, University of the Witwatersrand,(1978).

30. Adams G.R., Jager A.J. and Roering C. Investigations of rock fracture around deep-level gold mine stopes. Proceedings 22nd U.S. Symp. on Rock Mechanics. M.I.T. (1981) p.213-218.

31. Legge N.B. Rock dilatation ahead of deep level gold mine stope faces. Ph.D. Thesis. In preparation. University College Cardiff.

32. Pattrick K.W. The development of a data acquisition and pre-processing system for microseismic research. M.Sc. Thesis in preparation. University of the Witwatersrand.

33. Gay N.C. and van der Heever P.K. In situ stresses in the Klerksdorp Gold Mining District, South Africa - a correlation between geological structure and seismicity. Proceedings 23rd U.S. Symp. on Rock Mechanics A.I.M.E. Berkeley(1982) p. 176-182.

34. Roering C. Unpublished report (1978).

35. Roering C. and Rorke A.J. Rockbursts : Case Histories and Implications. In preparation.

36. Rorke A.J. and Roering C. Source mechanism studies of mining-induced seismic events in a deep level gold mine. Rockbursts and Seismicity in Mines. Editors Gay N.C. and Wainwright E.J. Symp. Series No. 6 S. Afr.Inst. Min. Metall. Johannesburg (1983)

37. Peirce A.P. and Ryder J.A. Extended boundary element methods in the modelling of brittle rock behaviour. Proceedings 5th Int. Soc. Rock Mech. International Congress on Rock Mechanics. Melbourne (1983) p. F159-F167.

38. Brummer R.K. Personal Communication.

Rock mechanics studies on the problem of ground control and rockbursts in the Kolar Gold Fields

R. Krishna Murthy M.Sc., M.I.M.M.
Mine Planning & Technical Services, Bharat Gold Mines, Kolar Gold Fields, Karnataka, India
P.D. Gupta B.Sc.(Min)., F.C.C.(Metal), M.M.G.I.(India)
Bharat Gold Mines, Kolar Gold Fields, Karnataka, India

SYNOPSIS

Two examples of major rockbursts resulting in large scale damages to underground workings and surface buildings & the rock mechanics investigations carried out over two & half decades which gave valuable information on the phenomenon of ground movement and rockbursts are discussed in the paper. New mining methods were evolved, reducing the frequency and intensity of rockbursts. The seismic network established on the field to locate the foci of rockbursts has helped in delineating zones of high stress and assessing stability of stoping excavations.

INTRODUCTION

Kolar Gold Field is situated at 12° 57' north and 78° 16' east in the southeast corner of Karnataka in India and lies at an altitude of 900 metres above mean sea level. There is indication that some of the workings may be more than 1000 years old. However, the modern phase of mining was started in 1880 and continued ever since. The problems of ground control and rockbursts associated with hard rock have been present in the Kolar Gold Fields since the beginning of this century.

They have occurred at all depths under different mining conditions. However the problems became serious as mining reached greater depths. One of the mines has reached depths of over 3200 m. Rockbursts have caused large scale fatalities, costly surface damages, loss of shafts, men travelling and haulage roadways, pumping and winding installations. Valuable proved reserves have been lost for ever. However, the problems of rockbursts have now been considerably reduced due to the introduction of better mining methods, based on rock mechanics studies.

Kolar Gold Fields.

Brief Geology & Stoping Methods.
The geology and mining practice of K.G.F. have been discussed in detail by Pryor[1] and Taylor.[2] The deposits occur in a belt of hornblende schist of lower Dharwar age, which is surrounded by granite and gneisses. The lodes dip towards west at 40° to 45° near the surface, and become vertical in depth. Of the many quartz lodes only two, the Champion and Oriental lodes are of great economic importance. The width of the lodes vary from 1m. to 6m. which have been interrupted by a series of faults, pegmatites and dykes. The Champion lode has been extensively worked out in all the three mines - the Mysore, Champion and Nundydroog mines which are operating at present (Fig.1), while the Oriental lode is being mined on a large scale only in Nundydroog mine due to economical reasons. In the Champion reef mine, the

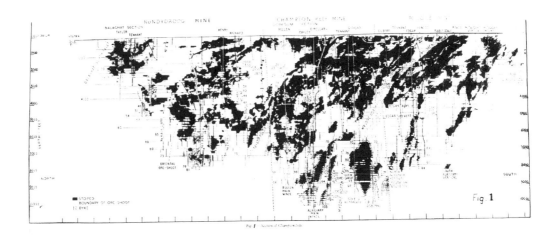

Fig. 1

main workings are on Champion lode at depths of over 3200 metres.

The extraction of lode is done mainly by (i) bottom stoping with granite support, (ii) stope driving with concrete and (iii) flatback stoping with sand fill. The rill system of stoping which was practiced at depth in Champion reef mine earlier has been replaced by stope driving.

Rockbursts

According to early records,[3] the first rockburst is reported to have occurred in a stope below 960 ft. level in the Oorgaum mine now a part of Champion reef mine in the year 1898. The rockbursts were classified in the early days as "air-blasts" and "quakes" depending on their intensity and area of damage.

At shallow depths, these problems were not critical except while mining shaft pillars and in exceptional cases Oreshoots which were highly stressed due to juxta-position of faults. However, they became serious as mining reached greater depths particularly when the Orebody to be mined was associated with faults, pegmatites & dykes, all involving planes of weakness. Large rich Oreshoots have been completely damaged and rendered unproductive as a result of severe rockbursts and in these cases, the first rockburst tremor was followed by a series of tremors over a period of several days. Buildings on

surface with 2-3 km. from the epicentral region have been damaged. The intensity of some major rockburst tremors was in the range of 4.5 to 5.0 on Richter scale and they have been recorded by the Seism-ographs located as far as 760 km. away from K.G.F. Fig. 2 shows the typical damages to steel setted drives as a result of rockburst.

Fig. 2 Drive after Rockburst.

Of these, the rockbursts that occurred on 27th November 1962 between 85 and 107 levels of Glen Oreshoot, and on 25th December 1966 below 97 level of Northern folds area of Champion reef mine, are discussed.

(i) Rockburst on 27th Nov.1962 (Fig.3)

The Glen Oreshoot which is a large Orebody on the Champion lode system in

Champion reef mine extends from 68 level to the bottom of the mine at 110 level & is about 400 metres long on strike. The reef is about 1 metre wide and dipping 84° to the west. A major fault known as "Mysore North fault" striking NNW converges on the Oreshoot especially on the north wing. The approach to the stoping area is by two major vertical shafts; one on the north side known as Heathcote shaft and the other on the south side known as Osborne shaft, and by crosscuts and footwall drives.

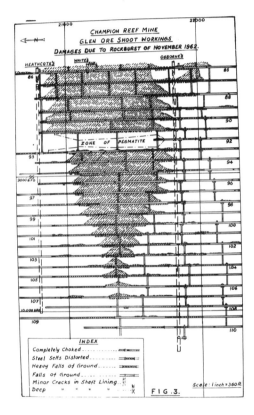

The Glen Oreshoot below 84 level of Champion reef mine was being mined by the rill system of stoping with a 'V' sequence, with dry granite walling as support, the rill stope face above the level being kept in advance of the face (saw-tooth) below the level and the level interval being 23m. & 30m. The reef drives and the portions of crosscuts and footwall drives where necessary were supported by steel arches, lagged with casuarina poles and filled with waste rocks.

The Glen Oreshoot is known to be rock-burst prone, particularly its Northern

wing. Typically, the damages are; distortion of steel arches, partial to complete choke of levels in reef drives and abutments, heavy falls of ground in the stope with partial to complete choke. Damages in foot-wall drives and crosscuts are usually minor with falls of ground in the unsetted portion of levels.

As many as 98 rockbursts occurred in Glen Oreshoot during the period 1942-62. The rockbursts occurring in the area may be generally classified into two categories; one where the damage is confined to the stope and stope abutments and the other to the footwall drives and complementary crosscuts east of Mysore North fault. Eight bursts belonged to the second category. The details of rock-bursts maintained in a systematic manner for the period 1957 to 1962 reveal that the frequency of rockbursts of minor & medium intensity is one in every 58 days while those of major intensity, one in 246 days.

The rockburst that occurred on 27th November 1962 in Glen Oreshoot between 85 and 107 levels damaged an area approx. 450m. in depth and 300m. on strike. This was the first of its kind in the history of mining in Kolar Gold Fields. Buildings on surface in the Champion reef mine area were also damaged. The first tremor was followed by a series of tremors; some of equal intensity which continued for several days. 59 tremors were recorded on the Weichert seismograph during the following 24 hours. Extensive damages had occurred to the 'sawtooth' abutments, indicating their complete failure. Heavy damages had also occurred in the cross-cuts immediately opposite the stope faces. The crosscuts and footwall drives opposite the stoped out ground was least affected. Heavy falls of ground occurred mostly in the unsetted portions of the shaft crosscuts, complementary crosscuts and footwall drives, lying remote from stope faces. Generally the damages appeared to be exentuated in planes of weakness; faults, calcite veins, pegmatite intrusions etc, occurring probably as a result of the shock waves emanated by the burst.

The major area collapse may be considered as a combination of the two categories mentioned earlier with utmost severity. There is reason to believe that there was a general seismic activity on the field during November 1962 as rockbursts resulting in damages were reported from the neighbouring mines also.

Examinations of the area between 90 & 93 levels, did not reveal, that the pegmatite barrier could have been the contributory factor for the occurrence of the rockbursts. The Mysore North fault is not considered responsible for the occurrence of the major movement by observations and measurements. It may have been responsible for accentuating the damages as they provide the plane of weakness. It would appear that the stoping may have reached a critical area of extraction, the abutment of stopes getting heavily loaded. The 'V' system of stoping with its sawtooth faces and the headings in the abutments at intervals of 23m.& 30m. may have formed into innumerable zones of high stress concentrations all along the longwall face. The whole longwall system was supported on the bottom of 'V' and naturally any major ground movement especially at the bottom of 'V' should throw the whole area into a state of unstable equilibrium resulting in severe damages. The rigid granite supports did not in any way help in controlling the severity of bursts.

It is very interesting to note that arching in crosscuts and cracking which are the indications of general movement of wall rocks were noticed throughout the Oreshoot for a few days before the major rockburst and particularly after the medium burst in 96 level North wing on 2nd November 1962. Two tremors were

recorded at K.G.F. Observatory before the major burst at 0220 hours on 27th November; one at 0114 hours and the other at 0219 hours. It could be that the burst at 0219 hours occurred at the bottom of 'V' and triggered off the major collapse. As already pointed out, there was a large scale ground activity going on the field and it is possible that it may have assisted in bringing about the major collapse in the already vulnerable area.

(ii) Rockburst on 25th Dec.1966 (Fig.4).

The Northern folds are a part of the Champion lode system. The folded formation of the quartz reefs which commences at the 73rd level has been developed to the 113th level, 3200 metres below field datum. The folds have an average strike length of 60 metres, and in plan view have a 'Z' shape; west limb, east limb and main reef which tends to elongate in depth where the average strike length increases to over 120 metres. The general pitch of the folds is northwards. The Mysore North fault, which is a major geological feature, lies only a short distance into the hanging wall of the folds.

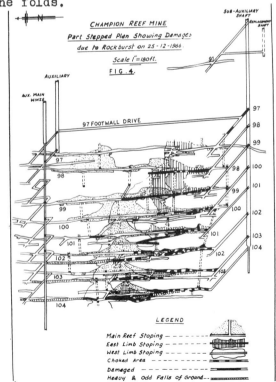

CHAMPION REEF MINE
Part Stepped Plan Showing Damages due to Rockburst on 25-12-1966.
Scale 1"=180ft.
FIG. 4.

LEGEND

Main Reef Stoping ————
East Limb Stoping ———————
West Limb Stoping ————————
Choked Area
Damaged ————————————
Heavy & Odd Falls of Ground——

Access to the area is by two major shafts i.e. Biddick's shaft on the south side and Auxiliary main winze on the north side of the stoping, the distance between the shafts being approx. 300 m. As Biddick's shaft got damaged frequently due to rockbursts, access to the stoping area is provided by two replacement shafts, one from 89 to 93 levels and the other from 93 to 103 levels.

In the Northern folds, extraction of the reefs was done on all the limbs by means of a rill system of stoping, using granite as the waste support. Above the 87th level, the three limbs of the northern folds were mined concurrently, with the main reef in advance of the east limb and west limb. The main objective of this system of extraction was that the stoped out ground on the main reef would provide a cushion for the shaft against the stresses induced by the east and west limb stopes.

However, by 1948, rockbursts affecting Biddick's shaft had become a matter of serious concern and it was decided to reverse the sequence of mining by taking the west limb first followed by the east limb, with the main reef being extracted last. This, it was expected, would initially destress the ground adjacent to the Mysore North fault and prevent the east limb stoping and the subsequent main reef extraction to build up progressively larger and more dangerous stresses against this barrier.

The main reef stopes had consistently given more trouble than either the east or west limb stopes and it was hoped, therefore, that the new sequence would improve matters by deferring the extraction of the main reef until the ground had been partially 'destressed' by the stoping on the other two reefs. This belief was also belied by the occurrence of major rockburst in December, 1966.

The major rockburst that occurred on 25th December 1966, affected the ground below the 97th level on the Northern folds. The first major tremor which occurred at 0807 hours was followed by 9 tremors within the next 45 minutes with as many as 65 tremors occurring during the week ending 31st December 1966. This is the first large scale area collapse that was experienced in this area. The damage in the stoping area was confined mainly to the abutments, the reef drives being severely damaged. Falls of loose occurred in the unsetted portion of the drives and crosscuts. From a visual examination, the damage appeared to be more marked below the 100th level. The condition of the stopes could not be assessed as entry was either partially or completely restricted in most places but the inspection carried out later revealed severe damages. From an examination of the 100th level, it was anticipated that the chokes on the south side of the stoping extended from the stopes to the sub-auxiliary shaft crosscut junctions. The damage below the 105th level could not be assessed as the area was under water. However, after de-watering the levels below 105 level, it was found that the damages to reef drives and stopes was as extensive as it was in the stoping area above 105 level.

The shaft lining of sub-auxiliary shaft between 97 and 98 levels suffered serious damage on the hanging wall side of the shaft. Severe damage to the 97plat of the shaft also occurred.

It is interesting to note that the bottom section of Champion reef mine was flooded due to unprecedented rains during preceeding 3 months of the burst. Large number of rockburst tremors were recorded on the field particularly during these 3 months.

The Northern folds were particularly susceptible to rockbursts and as many as 69 rockbursts occurred during the period from 1943 to 1966. The shaft was damaged on 14 occasions, the damage being restricted to the shaft only without any noticeable damages to the stoping area. The damage to the shaft has been confined mainly to the east side of the lining. Exceptions to this rule have been sink bursts which occurred during shaft sinking and the major area rockbursts that occurred on 25th December 1966. In this last case, the damage was mainly confined to west side of the lining. On 55 occasions the stoping area was affected. The frequency of rockbursts of minor and medium intensity was one in 84 days and of major intensity; one in 421 days according to the details of rockburst records maintained systematically for the period 1958 - 1966.

In this stoping area, a typical rockburst was an abutment failure just ahead of the stope face, similar to those failures experienced in the Glen Oreshoot, where rill stoping was also the standard mining method adopted. As a rule, the damage was not confined to any particular limb but extended to one, two or all of the limbs. A particularly vulnerable zone was the partition of black rock between the main reef and the east limb near the southern junction. Falls of loose rock often occurred in the unsetted portion of the reef drives, the crosscuts to the west limb from the main reef, and sub-auxiliary shaft crosscuts.

It was concluded that the burst may have occurred due to:

i) Stoping in a major folding system at depth having inadequate separation between individual limbs.

ii) Inhomogeneity of the dry granite support system with a void as high as 34% and the inability to control the immediate footwall and hangingwall of stopes especially at the junctions of east and west limbs where the reef width exceeded 7 m.

iii) The unstable geometric shape of the excavation with the formation of saw-tooth abutments at depth intervals of

23 m. & 30 m, as was in the case of Glen Oreshoot.

iv) Size of the excavations reaching probably the critical area of extraction (vertical span to horizontal).

The flooding of the mine in an area vulnerable to rockbursts such as folds may have contributed for the occurrence of the rockburst.

Rock Mechanics Research in Kolar Gold Fields

Investigations into the problems of ground control and rockbursts in K.G.F. was done in the past by a few individuals who were working in the field and by the 1926 and 1955 special committees set up for the purpose. The report of the 1955 committee[3] contains comprehensive recommendations on safe mining practice. The incidence and severity of rockbursts and accident to workmen have decreased as a result of adopting the recommendations of the committee as standard mining practice. However, a systematic investigation into these problems was commenced in 1955 when a rockburst research unit was formed and continued eversince. The work upto 1972 was done in collaboration with the University of Newcastle-Upon-Tyne, U.K., under the guidance of Prof. E.L.J.Potts.

The investigations which were carried out consists of statistical analysis on the available data, laboratory tests on the physical and elastic properties of rocks and field measurements on the rock-mass movement occurring in and around underground excavations.

The statistical analysis of the records of the rockbursts has revealed the following: [4,9]

i) Only a small percentage of the recorded bursts were traceable in underground. This is so, even in the Gold mines of South Africa.

Analysis into the energy released by rockburst tremors indicated that the bulk of energy resulting from mining is released by a small number of bursts of large amplitude. Calculations also indicated that the total energy released from rockbursts that occurred in major Oreshoots should have destressed the whole of stoping area and there should not be sufficient energy left for the occurrences of rockbursts in such frequent intervals as they have been experienced. It would appear therefore, that as soon as some energy is released due to a rockburst, the build up of energy takes place in the stoping regions over a period of time, the stresses built up in the rockmass around the region serving as a reservoir.

ii) Frequency of reported rockbursts is minimum on Sundays and maximum on Fridays.

iii) There is a significant peak in the occurrence of rockbursts during the time of stope blasting or soon afterwards.

iv) There appears to be a specific relationship between rainfall and rockbursts.

v) Frequency and severity of rockbursts are not directly related to depth. A large number of rockbursts of medium & major intensity have also occurred at shallow depths. The important factors for the causes of rockbursts are the physical and elastic properties of rocks, insitu stress, size and shape of the excavation, and the inhomogeneity of the rock such as the existence of faults, pegmatites, dykes and calcite stringers, all involving plane of weakness either in themselves or at the contact.

vi) The MAJOR AREA ROCKBURST resulting in very severe damages to the underground workings and surface buildings have occurred on the field at approx. 10 year period intervals.

vii) The minor & medium rockbursts have invariably occurred as a result of resettlement of ground, while the major area collapses have occurred due to the failure of abutments, remnants and pillars and time strain effects. The situation

has aggravated in the regions with geological disturbances.

viii) In the case of major area rockbursts affecting large Oreshoots, recurrence of spitting and arching in the footwall drives and crosscuts has been experienced over a period of two to three weeks prior to the rockbursts.

Laboratory Investigations
The laboratory investigations conducted since 1957 has provided valuable information on the physical and elastic properties of K.G.F. rocks.[6,7]

The Kolar schist has the average uniaxial compressive strength and young's modulus of elasticity 3000 kg/sq.cm. & 7.9×10^5 kg/sq.cm., respectively and the poisson's ratio being 0.2. The reef quartz has a high breaking strength of the order of 4200 kg/sq.cm. The results of the tests justify defining Kolar schist as a transversely isotropic medium. The violence of rupture of KGF rocks are found to be 2 to 3 times higher than those of South African quartzite.

Underground Investigations
A number of stress/strain measurements in and around rill stopes at depth including a major instrumentation around a pair of rill stopes in Glen Oreshoot, and a Flatback stope in the west reef, Nundydroog mine were carried out to study the rockmass movement pattern and stress build up occurring due to mining under such conditions. Details of earlier investigations has been discussed in the IMM Transactions[5] The important findings arising out of the investigations are:

In the Kolar Gold Fields, higher lateral stresses than can be accounted for by superincumbent weight alone exists in the virgin rock. In some cases, it is found to be greater than the vertical stress field and the ratio of the horizontal to vertical stress field varied from 1.6 to 4.

(a) Champion Reef Mine

In and around rill stopes with dry granite support at depth of 2315 m. - 2758 m.[7,8,9] (Fig.3).

i) Displacement pattern parallel to the plane of reef indicated an extensional zone in advance of the face and compressional zone behind the face bearing a strong similarity to the pattern observed above excavations in horizontal seams.

ii) At normal to the plane of reef, the measurements showed an erratic compressional movement in the foot and hanging walls of reef in advance of the face line and this is indicative of an unstable equilibrium existing there. Immediately behind the face and opposite the stoped out ground, the wall rock dilated. Further behind, there was recompression.

iii) Closure between stope walls behind the stope face was generally erratic and the maximum measured closure did not exceed 12-15% of stoping width though the voidage in the support was 34%. In some cases, significant rate of closure continued as far as 50 m. behind the face indicating that full consolidation of support is not attained for a considerable distance behind the face. These indicated a state of instability over a considerable distance behind the face.

iv) At the reef plane, there was indication of a "downward" movement of the hangingwall relative to the footwall in places behind the face. In advance of the face, the footwall rock generally "slumped" relative to that of hangingwall. Slumping of the footwall rock has been a marked feature during periods of rockbursts.

v) Stress measurements in advance of stope faces indicated the existance of very high front abutment stresses at distance varying from 1.2 to 1.8 metres in advance of the face. The maximum load measured in the granite packwall of a rill stope at a distance of 3.5 m. behind the

face was much lower than expected.

It was concluded that the solid granite supports used in this mine at depth and the slow rate of extraction (both of which have a major and adverse effect on mining economics) did nothing to assist in providing controlled relief to the wall rock and in minimising rockbursts. On the other hand, they assisted in building high stresses on to the solid abutments in advance of the faces and created a situation which became beyond control. Hence, it was found absolutely necessary to change the rill system of mining to a new system which would result in better ground control and the recommendations were as follows:

a) adopt a technique of geometrically designed sequence which permits a considerably increased rate of advance to be achieved and maintained; limit the area to be exposed after each machine cycle to the minimum and adopt a quick support system,

b) adopt a support system which yields and provides relief to the rockmass but is sufficiently strong to support the immediate walls, thus deliberately assist the main body of the rockmass to attain static and dynamic equilibrium quickly.

A system of "stope drive" with concrete supports for stoping rockburst prone Oreshoots and pillars at shallow depths, generally satisfied the above conditions.

Hence, the Northern fold area below the 97 level which was severely damaged due to rockbursts on 25th December 1966 was reopened and the extraction was commenced by the stope drive system with three longwall faces i.e. between 100 & 103 levels, 103 & 105 levels, and 105 & 109 levels (Fig.5). The extraction which was first restricted to east limb is now extended to main reef between the same levels with a lag & lead, between longwall faces. Since the commencement of stoping in 1971, only 5 rockbursts were reported from this area, as against 37

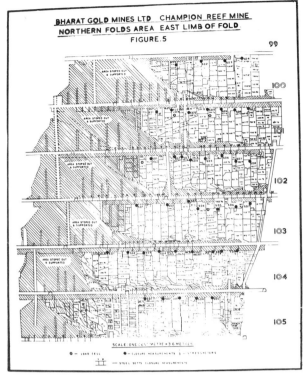

BHARAT GOLD MINES LTD   CHAMPION REEF MINE
NORTHERN FOLDS AREA EAST LIMB OF FOLD
FIGURE.5

reported for a period of 8 years between 1958-66 when stoping was being done by rill system of stoping.

Similarly, the stope drive system was adopted to extract the area in Glen Oreshoot which was affected by rockbursts that occurred on 27th November 1962. The approaches to the Oreshoot were reopened

and the rill promontories were shaped & a long wall system of stoping was adopted between 98 & 103 levels, on south wing & this was next followed on the north wing. The sequence of stoping is maintained in such a way that the stope faces below the level lie always one stope length in advance of stope face immediately above the level. Considerable amount of stoping has already been carried out with very little ground control problems (Fig.6).

The measured closures in stopes with concrete support is found to be more regular and uniform with advance of stope face and stability behind the face is reached much earlier as compared to those in stopes supported by granite walling. The average total closure between walls in the stopes supported by concrete is approx. 8 to 10 percent of stoping width over a face advance of 36.5 m., closure as much as 12 to 15% of stoping width is reached for distance of over 50 m. behind the face.

It may be interesting to note that during the period of 10 years prior to

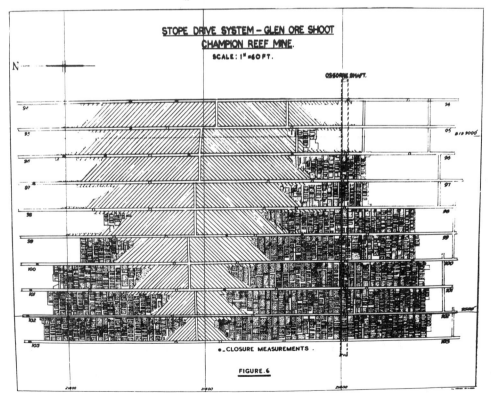

STOPE DRIVE SYSTEM - GLEN ORE SHOOT
CHAMPION REEF MINE.
SCALE: 1" =60 FT.

●_CLOSURE MEASUREMENTS .

FIGURE.6

the major rockbursts on 27th Nov. 1962, 58 rockbursts of major to minor intensity occurred resulting in damages to the workings and in most cases, the damages were very severe, while for the same period of 10 years since the commencement of longwall stoping in 1971 by the stope drive system, only 8 rockbursts occurred resulting in medium and minor damages.

(b) Nundydroog Mine

In this mine, the stoping is concentrated on Oriental lode. Geologically this lode is quite different from the Champion lode and is commonly intersected by NE-striking faults and by dykes striking east-west. Extraction is done by a flat-back system with hydraulic sand fill as support and longwall sequence of stoping, upto 62 level, 1750 m. below surface.

In this lode as many as 94 rockbursts were reported during the period 1957-72, nearly 41 percent occurring during the period when the stope was holing through to the level above. The damages due to rockbursts in the workings were found to be different from those experienced in the main reef stopes of Champion reef mine. Of these, the major rockburst occurred on 21st November 1971 resulting in damages to the stoping area from 3650 to 6200 levels over a strike length of 400 m. with significant damages to buildings on surface for over 3 sq.km. The first major tremor was followed by a series of tremors at short intervals and the bursting continued for several days. This was followed by another major rockburst in December 1972 damaging the workings between 5000 & 5900 levels over a strike length of 300 m.

Large scale instrumentation was done in and around the flat-back stopes to study the rockmass movement and stress build up due to mining.[10] A few important findings of the investigation are briefly mentioned:

i) The movement pattern around sand

filled back stopes is generally somewhat similar to what was observed around stopes in deeper levels of Champion reef mine but for some finer details such as area of compression and tensional zones, and magnitude etc.

ii) Closure between stope walls was more uniform approx. 10 percent of stoping width, though the measured voidage in the fill is 38 percent. It would appear that adequate relaxation of the strata around the stopes does not take place due to nature of wall rocks.

iii) Very high abutment stresses were measured within 0.5 to 1.5 metres from the face and in the sand fill, the maximum load measured was found to be rather low.

Closure Between Stope Walls

The analysis of wall closure measurements in stopes have indicated a correlation between rockburst and rate of closure on many occasions. In a few instances, there was a significant increase in the rate of closure varying from 2 to 16 times the normal rate for a few days prior to the rockbursts showing a 'V' phenomenon while in others, there was no such indication. In some cases, the increase in rate of closure dropped and any subsequent sudden increase was followed by a rockburst.[11,12] It is interesting to observe that strain measurements in crosscuts carried out in Glen Oreshoot, indicated a somewhat 'U' phenomenon, a "contraction" followed by an "expansion" during periods when rockbursts occurred in the area.[5,3]

Statistical analysis of results also confirmed the existence of a significant difference between periods and between bays before the occurrence of some rockbursts. The sudden large closure which occurred as a result of the burst was followed by a gradual decrease in the rate of closure varying from three days to two weeks after the bursts.

As the closure measurements are found to be very useful to assess the day-to-day stability of underground workings it has now become a standard practice to introduce these measurements in the areas vulnerable to rockbursts in spite of their limitations.

SEISMIC INVESTIGATIONS

A break through has been made in the field of rockburst research in Kolar Gold Fields by the application of seismic techniques to study the stability of underground workings.

Monitoring of seismic activity in stoping areas was done in Kolar Gold Fields for a short duration during 1968-69 in collaboration with the University of Newcastle-upon-tyne, U.K.[13] It was concluded from the experiment that the technique could be successfully adopted under KGF mining conditions. Encouraged by the above experiment a seismic network consisting of 15 geophones (8 on surface and 7 underground) with their electronics was established to cover an area of 6 km. x 3 km. on the field, (Fig.7). The signals picked up by the geophones located close to the recording station are directly transmitted through over-head cables while from those located at greater distances from the laboratory are transmitted through cable driven PPM, and recorded on a 24 channel magnetic tape recorder/replay system. The recorded signals are transcribed using a mingograph recorder and from the arrival times of signals (P-wave) at various geophones, the foci of rockbursts are computed using a micro-processor. The frequency coverage for the direct channels was DC to 100 Hz. frequency modulated (FM) telemetry on a carrier frequency of 540 Hz. while that of PPM channels was from DC to 45 Hz. Due to low frequency limitations recorded by PPM channels, the individual inputs of geophones were also directly telemetered by cables to increase the frequency coverage and the system response has now been increased to 180 Hz. for better accuracy in the detection of the arrival time of the signals. The accuracy of reading the onset of signal is $\pm$ 2 ms. The block diagram of the system is shown in figure 8 and recording instruments are shown in figure 9.

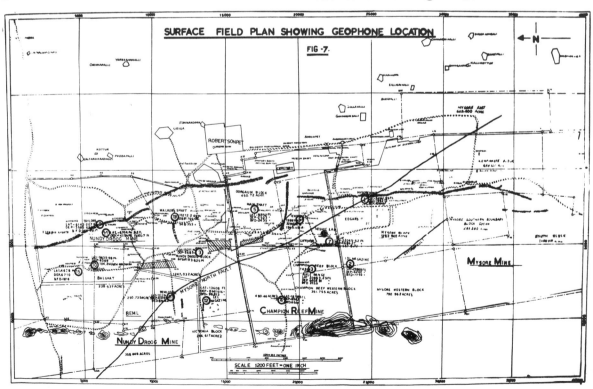

SURFACE FIELD PLAN SHOWING GEOPHONE LOCATION

FIG -7.

SCALE 1200 FEET = ONE INCH

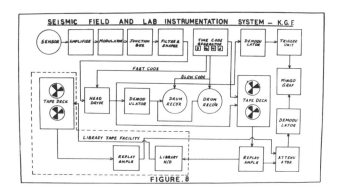

FIGURE.8

Fig·9

The sensors are Electromagnetic type, manufactured by GEO SPACE, the output sensitivity being 0.33 v/cm/sec. at 0.6 critical damping. The signal cable is of special make; 4-core polythene sheathed GI wire braided; mutual capacitance 0.055 F/km and attenuation 1db/km at 1Khz. Data storage is 1" wide 24 track magnetic tape, tape speed; 0.625 inch/sec., Electronic gain: 80 db, Total magnification at 1v/cm. at 7.5 Hz: 1,00,000, Dynamic range : 40 db.

The geophone network was established in Kolar Gold Fields during 1978 and the seismic activity occurring around mine workings have been continuously monitored since September 1978. The work was implemented in collaboration with Bhabha Atomic Research Centre, Bombay.[14]

Between the period from September 1978 to 31st December 1981 as many as 3,483 rockburst tremors of those intensities which could be picked up by five or more geophones in the network were recorded. Of these, 51 rockbursts resulting in

damages to underground workings were reported from the mines. The co-ordinates of the observed damages in the case of these rockbursts have also been used to deduce a fairly consistent picture of the velocity structure in the KGF region. A few blasting experiments which were conducted in some areas in the mines as an R&D effort to improve production, was used to improve the velocity models for those regions.

In the first case, assuming the rockmass as homogeneous isotropic medium and from the data of known rockbursts and suitable combination of geophones, three velocity models for the KGF region were arrived at; P-wave velocity of 6.97 km/sec. for Nundydroog region, 6.94 km/sec. for the region between Nundydroog and Champion reef mines and 6.53 km/sec. for the Champion reef and Mysore Mines.

In the second case, the rockmass is considered "anisotropic". From the data of known rockbursts and choosing suitable combination of geophones, different sets of regional velocities were computed:

Area	Velocities in km/sec.		
	V_x	V_y	V_z
Northern Folds (east limb)	6.2	5.3	6.6
Northern Folds (North of dyke)	6.1	5.4	6.2
Glen Oreshoot (north wing)	6.2	5.4	6.6
Glen Oreshoot (south wing)	6.2	5.4	6.3
Nundydroog mine	6.6	6.45	6.5

The co-ordinates of computed foci using the regional velocities have agreed within ± 25 metres of the locations of damages. Wherever there was any appreciable difference in the co-ordinates between the computed foci and the locations of damages at a particular region, improvements were made by trial and error to obtain a better fit and the most suitable velocities are considered for computation of foci in the event of a subsequent rockburst. The computed foci of rockbursts recorded from the bottom section of Champion reef mine is shown in Fig. 10.

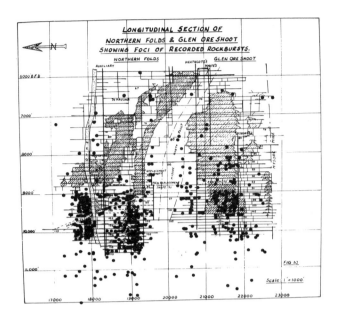

On many occasions work has been suspended and men withdrawn depending on the level of seismic activity indicated by the network and the strain measurements.

MICRO SEISMIC INVESTIGATIONS

Encouraged by the results of the above investigations, instrumentation is being developed in collaboration with BARC to monitor the micro-seismic activities in Glen Oreshoot with a view to forecast their occurrences.

The instrumentation consists of a close network of high frequency sensors with power supply, amplifier, highpass filter and modulator. The rock noises picked up by the sensor network are transmitted from underground to surface through special cables to the laboratory located on surface, and demodulated. A delay processor in the system measures the lags of signal arrivals at different sensors and a micro processor computes the locations and prints out the required information such as foci, rate of rock emission etc. A computer will also be installed on-line for detailed analysis.

The frequency coverage is upto 5 KHz and the accuracy of signal detection is ± 0.1 ms.

Conclusions

From the above investigations it has been possible to obtain useful information on the stress/strain distribution in and around underground workings and on the strata displacement characteristic of rockbursts. This has been of great assistance to plan improved mining methods in areas vulnerable to rockbursts especially in the bottom section of Champion reef mine and to exercise a better control over them. The seismic monitoring of rockbursts has become such a "handy tool" for assessment of safety of mine workings that the work is either suspended or resumed on many occasions, depending on the seismicity recorded from rockburst prone areas.

Acknowledgement.

The authors are grateful to Prof. Potts for his inspiration and guidance in conducting investigations carried out upto 1971, to authorities of Bhabha atomic research centre, Bombay, without whose help the seismic investigation would not have been possible and Department of Mines, Ministry of Steel and Mines, Government of India for encouragement and funds to conduct seismic and micro seismic investigations.

References.

1. Pryor T. The underground geology of the Kolar Gold Fields. Trans. Instn Min. Metall., London, Vol.33, 1923-24, P.95-115.
2. Taylor J.T.M. Mining practice on the Kolar Gold Fields, India. Trans. Instn Min. Metall., London, Vol.70, 1960-61, P. 575-604.
3. Messrs. John Taylor and Sons Limited. Report of the special committee on the occurrence of rockbursts in the mines of the Kolar Gold Fields, Mysore State, India, 1955.

4. Issacson E.De.St.Q. A statistical analysis of rockbursts on the Kolar Gold Fields. Bull. KGF Min.Metall.Soc. Vol. 16, P. 85-103, 1957.

5. Taylor J.T.M. Research on ground control and rockbursts on the Kolar Gold Fields, India. Trans. Instn Min. Metall., London (1962-63); 72: (317 - 338).

6. Bhattacharyya A.K. Investigations into the elastic and strength properties of Hornblende schists from the Kolar Gold Fields. M.Sc. Dissertation, University of Durham - 1962.

7. Miller E. A study of rock pressure and movements about a nearly vertical reef at great depth with particular reference to rockbursts and design of mine openings. P.hd. Thesis, University of Durham - 1965.

8. Miller E. Notes on rockmechanics research in the K.G.F. KGF Min. Metall. Soc. Bull. Vol. 95 P.23-83.

9. Krishna Murthy R. A review of rock-burst research in the Kolar Gold Fields-proc.of symp. on rockmechanics; Publ. by Min. Metall. Div. of the Instn. Engrs. (India) July 1972.

10. Sibson J.N.S. Investigations into design and stability of workings in a deep nearly vertical reef subject to rockbursts. P.hd. Thesis, Dept. Min. Engr. Unive. Newcastle-Upon-Tyne, 1967.

11. Krishna Murthy R. Strata displace-ment around a vertical shaft due to the mining of a protective pillar in an intersecting highly inclined reef prone to rockburst. M.Sc. Dissertation, Univ. of Newcastle-Upon-Tyne, 1966.

12. Krishna Murthy R. & Nagarajan K.S. Strata control measurements in stopes. Golden Jub.Symp. Banaras Hindu Univ. 1976.

13. Bhattacharyya A.K. Application of seismic techniques to problem of rock-mechanics. P.hd. Thesis. Univ. of Newcastle-Upon-Tyne, 1967.

14. Murty G.S. & Krishna Murthy R. Seismic investigation of rockbursts in Kolar Gold Fields. Paper presented in Seminar on recent trends in Gold Mining Practice, Bharat Gold Mines Limited - December 1980 (To be published).

Rockbursts at Macassa mine and the Kirkland Lake mining area

John F. Cook B.Eng., C.Eng.(UK), P.Eng.(Ontario), M.I.M.M., M.C.I.M.
Lac Minerals, Ltd., Kirkland Lake, Ontario, Canada
Don Bruce M.C.I.M.
Lac Minerals, Ltd., Macassa Division, Kirkland Lake, Ontario, Canada

SYNOPSIS

Macassa Mine is situated at the western end of Kirkland Lake mining camp in Northern Ontario, Canada. The other mines in the area which worked the same orebody or series of orebodies closed several years ago. Macassa has been working since 1933 and, since 1965 has been the only operating mine in Kirkland Lake.

Current stoping operations at Macassa range from 2,300 to 6,500 feet below surface. Other mines in the area, particularly the Lake Shore and Wright Hargreaves mines went to greater depths, up to 8,200 feet below surface. The whole Kirkland Lake mining camp has a history of rockbursts stretching back over fifty years. Generally, the bursts in the Lake Shore and Wright Hargreaves Mines were the most severe. These mines had the greatest ore expectancy and extracted a large areal percentage of the reef. By contrast, mining at Macassa has been relatively scattered and the frequency and intensity of bursting has been correspondingly lower. The history of bursting and mining practices have been examined.

The western side of Macassa at depth has a significantly increased ore expectancy and more intense stoping operations are planned. This, coupled with a recent large rockburst, has caused the mine to re-examine methods to predict and prevent bursts in the future. The mine is relatively small and the emphasis has been on unsophisticated techniques. Considerable attention has been given to the variation in properties of the rocks that make up the orebody and the immediate wall rocks. Rockburst prone areas can be identified by measuring seismic velocities and detailed geological mapping to show areas of stiffer rock. Special precautions can be taken when stoping through these areas and de-stressing or rock softening is practiced.

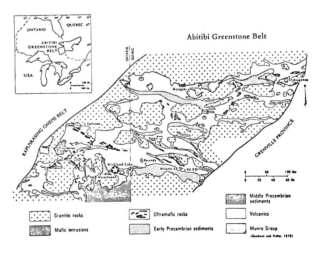

Fig. 1 Geology of the Abitibi "greenstone" belt and location of Kirkland Lake (geology after Goodwin and Ridler 1970) from Watson & Kerrich.

INTRODUCTION

The Kirkland Lake mining area is situated in Northern Ontario, Canada, (see Fig. 1). Gold ore has been mined in this area from steeply dipping narrow deposits for almost seventy years. These veins are all related to large regional fault zones or breaks as they are known locally.

Originally there were seven mines active in the area but only Macassa Mine is now operating. This mine is situated at the western end of the mining area, (see Fig. 2), and has now been operating for about fifty years. A new 7,275 feet deep shaft is currently being sunk to exploit reserves at the west end of the property.

The whole mining area has a history of rockbursts but at Macassa Mine this problem has not been severe. In the area to be exploited by the new shaft, the areal ore expectancy is significantly higher than elsewhere. Action to combat bursting has had to be increased.

This paper has been prepared to comment on the history of bursting in this particular mining field and to outline measures currently being taken or considered to avoid rockbursts in the future.

4 Macassa
5 Kirkland Minerals
6 Teck Hughes
7 Lake Shore
8 Wright Hargreaves
9 Sylvanite

Fig. 2 General Geological Map of Kirkland Lake Area.

GEOLOGY

The general geology of the region is shown in Figures 1 and 2 and the geology of the mine has been described in detail elsewhere [1]. The main rock types and structural changes are classifiable by age as shown below.

PERIOD -(All Pre-Cambrian)	ROCK TYPE	STRUCTURAL CHANGES
Post Algoman		Post ore faulting
	Quartz Diabase	
	Intrusive Contact	
Algoman		Gold deposition Wall rock alteration Pre ore faulting
	Syenite porphyry Quartz-feldspar porphyry Syenite Augite Syenite	
	Intrusive Contact	
		Folding
Timiskaming	Tuff Conglomerate/Greywacke	
	Unconformity	
Keewatin		Folding
	Intermediate to basic meta-volcanics	

The whole area is overlain with 0 - 50 feet of Pleistocene glacial deposits of clay, sand and fill.

The rock formations in Macassa Mine and the other Kirkland Lake mines consist of a variety of Algoman intrusives with Temiskaming sediments and tuffs. The sedimentary units and the long axis of all the intrusive bodies strike N 60 - 80° E and dip steeply to the south. Their dips flatten somewhat at depths greater than 4,000 feet. The sediments are most commonly found on the north and south flanks of the elongated composite stock. Augite syenite is the oldest and most wide spread of the intrusive types. Syenite porphyry occurs as sills with sharply defined intrusive contacts and cuts the syenite and augite syenite while conforming fairly closely in strike and dip to the regional trend of the formations. The composite stock dips steeply to the south, widens with depth and apparently has a steep overall pitch to the west. In recent years it has become apparent that this stock is narrowing considerably going west.

The main sequence of geological events in the Kirkland Lake camp has been; folding of the Temiskaming series; intrusion of the syenite and porphyry stock; faulting and fracturing; ore deposition into the breaks so formed; additional movement on breaks and gold re-mobilization; diabase dyke and minette dyke intrusion and post ore faulting.

The Temiskaming rocks form the north limb of a syncline and lie unconformably on the older Keewatin volcanics which occur north of the region. These folded sediments were invaded by the intrusive stock with a strike roughly parallel to the strike of sediments and a slightly steeper dip. The principal gold bearing veins of Kirkland Lake occur along one of several fault systems which has been designated as the Main Break on Kirkland Lake Fault. The Macassa Mine workings are on the westerly extensions of this fault system and have been referred to as the '04' Break and the South Break in underground mine nomenclature.

The Main Break traverses the entire length of the Macassa Property and the whole Kirkland Lake camp as a clean cut fault marked by a zone of mylonized and sometimes brecciated wallrock which is normally rehealed. It is a thrust fault with an estimated displacement of approximately 1,500 feet and dips steeply to the south. The deeper levels of the mine exhibit a branching fault system with the development of a split to the west, forming a north and south branch to the Main Break. In addition, another sub-parallel fault of major importance is connected to the north branch of the Main Break via cross breaks.

Post ore faulting is characterized by a few relatively major faults which can cause locally poor ground conditions, apart from causing orebody displacement.

The gold bearing veins occur in quartz filled fractures associated with the pre-ore faults and fractures. Subsidiary veins are well defined quartz filled fissures, whereas along the major faults there are variations from single fissure veins to intricately connected composite veins or lodes with the distribution of quartz being very spotty and irregular.

The present average ore body width varies from 7 to 8 feet. Wider zones do occur and may increase in the future. Figures 3 and 4 show a surface geological plan and section respectively.

Within the area of the ore and its immediate sidewalls, the rock types vary significantly which in turn causes large variations in mechanical properties. There is an interleafing of

porphyry, basic syenite and tuff. In addition, all these rock types contain varying quantities of quartz with a concentration in the orebody.

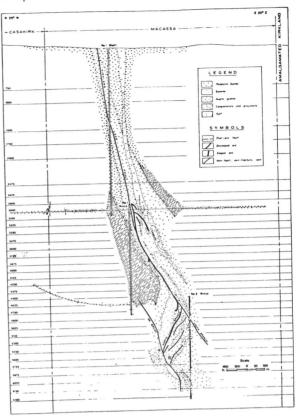

Fig. 3. Macassa Mine, geological section through No.1 Shaft,(after Charlewood 1964) from Watson & Kerrich.

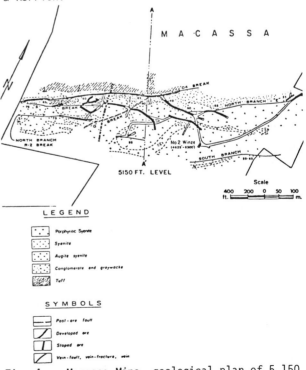

Fig. 4. Macassa Mine, geological plan of 5,150 ft. level,(after Charleswood 1964) from Watson & Kerrich.

The strengths and properties of the various rock types have been determined from a small number of samples. The following conclusions can be drawn:

Rock Type	Uniaxial Compressive Strength psi	Triaxial Compressive Strength with 2,000 psi confining pressure psi	Modulus of Elasticity 10^6 psi
Porphyry	33,000	56,000	10
Tuff	28,000	56,000	12
Basic Syenite	21,000	46,000	9 - 30

Very few measurements of the modulus of elasticity have been made in the laboratory but generally:
- The tuff is stiffer than the porphyry which is stiffer than the basic syenite.
- The basic syenite can be much stiffer than the other rocks possibly as a result of increased silica content.

Limited seismic work in 1978 indicated wave velocities in the orebody ranging from 7,000 to 23,000 ft/sec. This indicated a variation in bulk elastic modulus from 2 to 20 X 10^6 psi over very short distances.

More recent work has confirmed these findings.

The in-situ stress environment at Macassa and the whole Kirkland Lake camp is not known but it is highly probable that the horizontal stress normal to the orebody is at least equal to the vertical stress and possibly significantly higher. The following stress conditions are assumed:

Vertical Stress   6,500 psi
Horizontal Stress
normal to orebody   6,500 to 13,000 psi

THE ROCKBURST PROBLEM - HISTORICAL
The first rockbursts in the camp were recorded in 1932 at the Lake Shore and the Wright Hargreaves mines. By 1942, the problem had grown significantly and R.G.K. Morrison visited the area from Kolar to examine the mines and advise the mine owners [2]. The subject was also reported in some detail by various authors in 1939 and 1940 and by W.T. Robson in 1946 when he was Superintendent at Lake Shore [3]. Both these researchers recognized the basic problems of the sequence of mining causing pillars which in turn led to high stresses and the possibility of bursting. Robson attached a great deal of importance to the geological structural weaknesses and,

to a certain extent, the variation in rock types. He envisaged movements of large scale blocks bounded by fractures which could locally cause large increases in stress. Geological weaknesses were also suggested as a direct mechanism of bursting. The following quotation from Robson's paper is still very relevant today:
" The state of stress is developed by forces which are attempting to produce slippage along the fracture plane but, since they are opposed by friction, energy is stored up in the rock. If the opposing forces do not remain in balance, a slip occurs on the fracture plane, the release of the stored-up energy being accompanied usually by a shock or tremor."

It was also noted that the presence of more schistose rock in the potential failure area allows energy to be dissipated peacefully. This was true even when large continuous areas of ore were mined at depth.

Bursting in the Lake Shore Mine in particular was more severe in areas which seemed to correlate with a more complex arrangement of geological weaknesses. There may have also been a correlation with variation in rock types. It was specifically noted that development was liable to small strain bursts (away from stoping) in certain areas which were probably correlatable with rock type.

The following description of bursting in two areas of the Lake Shore is of interest.
" Down to the 1,600 foot level, the general practice at Lake Shore, as at most other mines in this part of the country, was shrinkage stoping. With this method, the proper filling of the stopes (using dry sand fill) after they had been drawn empty was found to be very difficult, and the support afforded the walls by such incomplete filling, or even by the ore in shrinkage stopes which had not been drawn, was inadequate to prevent the continued sag of the hanging wall. Consequently, as the walls in the area had not reached a state of equilibrium, it is reasonable to suppose that the unstable condition influenced, in some degree, the breaking up of the walls of the workings below the 1,600 foot horizon.

Below the 1,600 foot level, horizontal cut-and-fill methods of mining were employed, with filling following closely on the mining of the ore. The stopes above the 1,800 foot and 2,000 foot levels were mined by this method without encountering any particular problem, but when those above the 2,000, 2,325, 2,450, and 2,575 foot horizons had advanced to within approximately 40 feet of the levels above, bursting occurred. At this time, the ore remaining to be mined in this area lay in a series of long horizontal floor-pillars of relatively limited depth. To complete the mining of this ore and to confine the bursts to the smallest possible areas, the sill floors were cut up by additional raises and the ore recovered on short rills. The result was the formation of a large number of isolated pillars which were gradually being reduced in size and number. Under these conditions, some of the pillars became over-stressed and numerous bursts occurred. The formation of several small pillars resulted first in the violent failure of individual pillars, and later, as mining advanced, in the disruption of floor pillars over distances up to 250 feet along the strike of the vein. Figure 5 shows, in plan and cross-section, the principal fractures occurring in this area which constitute lines of weakness in the hanging-wall of the orebody.

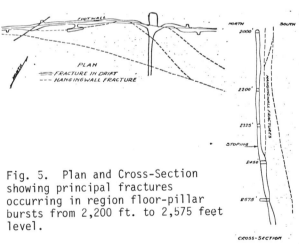

Fig. 5. Plan and Cross-Section showing principal fractures occurring in region floor-pillar bursts from 2,200 ft. to 2,575 feet level.

The horizontal slice, square-set, cut-and-fill method of mining, by means of which the footwall orebodies in the west half of the property were mined on the levels from the 2,000 foot to the 2,575 foot horizon resulted in the formation of horizontal floor-pillars of relatively shallow depth. Their recovery was a difficult problem, fraught with a high incidence

of rockbursting, as described above. Below the 2,575 foot horizon, the method of mining adopted for use in this part of the mine was a steep square-set rill system of stoping, employing close filling. At the same time, a planned sequence of stoping was also instituted, whereby stoping would start from each side of two vertical raises driven on each level at distances of 400 feet and 1,200 feet west of the main cross-cuts.

The sequence of stoping was carried out for several levels below the 2,575 foot horizon without undue trouble from bursting, and raises from the downward extension of the sequence were driven as far as the 3,825 foot level. As stoping reached greater depths, however, it was seen that the pillar formed by approaching stope faces in the middle of the ore shoot would eventually lead to a mining problem of considerable difficulty. It was decided, therefore, to limit further the distribution of stoping sections on the levels below the 3,825 foot horizon, and a sequence of stoping was adopted in which stoping started both ways from a vertical raise, sited at approximately the middle of the ore shoot, in place of the two raises formerly used. While this arrangement reduced drastically the rate of production from each level, it had a decided advantage in that no pillars were formed in the middle of the orebody. Figure 6 is a longitudinal section illustrating the arrangement of stoping sections in this part of the mine."

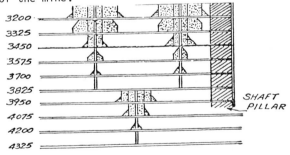

STOPING ARRANGEMENT
3075' TO 4325' HORIZON
Fig. 6

In 1964, F. Buckle, the General Manager of Wright Hargreaves, reported on rockbursts in that mine [4]. This author also emphasized the importance of variation in structure and rock type.

The Lake Shore and Wright Hargreaves mines

were characterized by higher areal ore expectancy
and large continuous areas of the reef were mined.
Values in the other properties were more
scattered and generally the rockburst situation
was less severe. In the two larger mines, long
wall type methods using rill cut-and-fill were
used to improve the sequence of mining. The
restriction of closure by fill was recognized as
a useful measure. This was more to restrict
movement along geological features, thus preventing a build-up of 'secondary weight', rather than
any energy considerations.

Where pillars were formed and needed to be
extracted, it was relatively common, in the 50's
and 60's, to use de-stressing techniques to
improve conditions. Practically, these techniques were considered to be useful and successful.
It is interesting to examine an example of de-
stressing quoted in a paper by J. Harling [5].

Figure 7 illustrates two stopes in the Lake
Shore Mine approaching the Lake Shore Fault. This
fault had contributed to the bursting hazard on
all horizons. This de-stressing exercise was
done in 1958 and was an early example of the
use of the technique. Later practice would have
used holes closer together, about 25 feet apart.
However, this application was successful and the
area was stoped without problem.

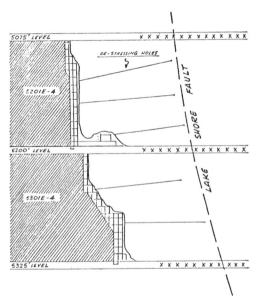

LONGITUDINAL SECTION, SOUTH VEIN, LAKE SHORE MINE,
SHOWING STOPES AT 5200-FOOT AND 5325-FOOT LEVELS
APPROACHING THE LAKE SHORE FAULT. THIS ILLUSTRATES
USE OF LONG HOLES TO DE-STRESS AN AREA BETWEEN
ADVANCING STOPE FACES AND A MAJOR FAULT.

▨▨▨ BACKFILLED
|x x x x| BLOCKED BY PREVIOUS BURST

Fig. 7

Mining at Macassa has always been more
scattered than at Lake Shore and Wright Hargreaves due to the distribution of values. Generally
the rockburst problem has been less severe but
bursts have occurred, particularly in pillar
areas. De-stressing has been used on a regular
basis in the past 20 - 25 years with apparent
success.

THE ROCKBURST PROBLEM - PRESENT
Macassa is now the only operating mine in the
Kirkland Lake camp and has been so since 1965.
The bulk of the ore has been extracted with
horizontal cut-and-fill methods. The area
now being developed has economic values over
a larger area of the break at depths of 5,000 -
7,000 feet on the west end of the mine. A
major new shaft is being sunk to exploit these
reserves. With the higher percentage extraction
that will result, greater attention must be
given to combatting rockbursts. This aspect was
emphasized by a relatively major rockburst that
took place in mid-1982 on the fringe of the
new mining area. This burst took place in an
area where high stresses and burst conditions
were not anticipated. It thus provided the spur
for further study. It is worth considering this
burst in detail and outlining the current work
and theories on the problem.

The burst occurred on the western extremity
of the mine (and of the camp). As a result of
the burst, damage occurred on 60 level, 61/38
sub-levels 1 and 2, 61/38 raise and 61/37 raise,
see Figure 8. The burst occurred immediately
after the detonation of a small number of holes
in the stope just above 61 level close to 61/37
raise. The event consisted of a large initial
burst, followed by several smaller shocks. In
this stope, an experimental mining method with
higher productivity was being tried. This is
essentially a long hole cut-and-fill method with
inclined slices. This method is normally used
to extract sill pillars only, with the bulk of
the stoping by horizontal cut-and-fill (using
waste rock fill).

The ore in this area is mainly contained in
a basic syenite, which also forms the hanging
wall and footwall, but tuff and porphyry are also
present. The quartz content of the orebody is

generally higher than the surrounding country rock but this had not been quantified at the time of the burst. However, during development of the block the crews noted relatively hard drilling conditions which is possibly indicative of high silica content in the orebody. They also noted an increase in 'spitting' or minor strain bursts.

caused by mining in the 61/37 and 61/38 area. This is because of its situation and because the higher modulus would attract higher stress.

- By being relatively stiff by comparison with the hanging wall and the footwall basic syenite, the pillar had the potential to fail explosively.

In view of the history of bursting at Macassa, it is surprising that the 61/38 burst

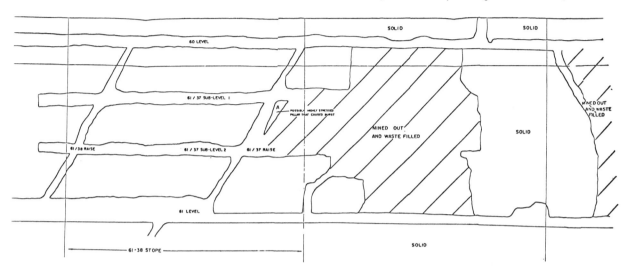

Fig. 8.   Longitudinal Section showing 61/38 Stope

The burst area is situated on the western edge of stoping and stress concentrations will only be those associated with the abutment rather than the very high stresses associated with pillars. The possible exception to this is the small pillar marked 'A' in Figure 8. The general stress level in this pillar, normal to the ore-body, could be about twice the general horizontal stress say, 26,000 psi with higher values at the edges, Stress levels at the edge of the stope would not be greater than around 20,000 psi.

As has been mentioned earlier, the modulus of rocks at Macassa varies widely and it is postulated that this variation plays a key role in the form of rock failure.

Immediately after this particular burst, the following mechanism was proposed:

Pillar 'A', which is in tuff, was the stiff structural member which caused the rockburst, because:

- The tuff is stiffer than the surrounding rock and therefore was more highly stressed before any openings were made.

- The pillar became the area of highest stress

was so large. It is surmised that the general stiffness of the orebody is high in this area and normal levels of peaceful failure had not occurred. Thus, when stresses reached failure at one particular place (Pillar A) there was a large quantity of stored energy available which was released suddenly. The resulting shock waves caused the damage which was observed.

Generally it is considered that the mining method(s) in use at Macassa are good from the point of view of minimizing closure and thus energy release. Apart from considerations of good sequential mining it was thought that areas of burst potential may be identifiable by examining geological and modulus variations in detail.

Seismic work is still proceeding but initial results show a strong correlation between velocity/modulus variations and detailed geology. Measured variations have ranged from 8,000 ft/sec. to 20,000 ft/sec. Increasing velocity is correlatable with increasing quartz content and the presence of tuff.

Measurements of velocities in the vicinity of the 61/38 rockburst showed wave velocities significantly lower than would be expected from the geology. It was concluded that fracturing caused by the burst had effectively softened this area.

Since the burst, the 61/38 area has been rehabilitated and the pillar which was thought to be the focal point of the burst was found completely intact. The initiation point remains unknown but mining is proceeding without any further incidence of bursting.

DISCUSSIONS

The 61/38 rockburst was the largest in Macassa's history and, whilst it did not cause any injuries, it caused considerable disruption to stoping. It took place in a relatively low stress environment on the edge of the area which will be the main source of ore when the new shaft is complete. This area is marked by very high ore expectancy along the break. Thus during stoping:
  - few waste pillars will be left;
  - stresses will be higher in pillars and at abutments;
  - closure will increase;
  - energy release will be higher.

It was therefore a cause for considerable concern.

Areas immediately to the east of the burst area had been stoped without blasting even though pillars led to higher stresses. The reason for the 61/38 burst is basically due to the geological geometry. Relatively large variations in stiffness in a small area led to differential stressing and strain energy distribution. This, in turn, led to an explosive fracture. This large variation in mechanical parameters had not been fully appreciated up till this point.

Geological conditions such as those encountered in 61/38 are expected to re-occur and seismic mapping is being undertaken to identify such areas. The question then remains as to the action required in these areas.

The method of mining will continue to use rock fill close to the working face, thus limiting closure.

The mining sequence will be designed so that stress concentrations are minimized with a constant rate of energy release.

Compression of the fill will continue to absorb energy. Despite this, bursts can still be expected in areas of unfavourable geology. If these areas have been identified, extra support can be installed so that, when bursts occur, damage can be minimized.

However, it is also of interest to investigate whether stiff areas can be softened. This would enable energy to be dissipated in a peaceful rather than a violent manner. It is postulated that softening can be achieved by the use of blasting de-stressing holes. De-stressing is really a misnomer as the main object of the exercise is to soften the rock so that it fails peacefully when subjected to stress.

There is considerable disagreement between South African practice and North American practice. In South Africa, it has been reported that de-stressing is not beneficial. This is because, during monitored de-stressing tests, no release of seismic energy was observed, neither was closure.

On the other hand, in North America, de-stressing has been carried out on a routine basis and it has appeared to reduce the incidence of rockbursts. Little attempt has been made to confirm scientifically that it has been the de-stressing that helped. There is one notable exception to this: the Galena Mine in Idaho [6]. In an experiment, de-stressing caused a softening of a remnant pillar and also caused closure, indicating that, in fact, real de-stressing of the pillar had taken place.

In the opinion of the writer, the Galena Mine example is an extreme case and the pillar concerned was fractured too intensely. The object of de-stressing blasting should be to increase the internal fracture surfaces and thus decrease the bulk modulus of elasticity. A reduction in stress and a manifestation of closure is not necessary to achieve this. The overall competency of the ground must be changed very little and any sort of general fracturing must be avoided. Otherwise one problem is merely transformed into another.

The 61/38 rockburst has provided some evidence that rock softening is possible. Measurements of wave velocity in the area have shown that the bulk modulus of the rock is significantly lower than would be expected from the geology. This cannot be confirmed by measurements as no wave velocity measurements were done prior to the burst. However, it was noted by miners to be hard to drill and prone to small strain bursts. After the broken rock that collapsed into the tunnels was cleaned up, the remaining rock was quite unfractured and stable. Work proceeded to drill holes and generally get the area back into production with few problems.

If a block of high modulus ground could be softened using long hole blasting, the likelihood of bursting could be reduced significantly. A de-stressing or softening experiment along these lines is planned. An area of high wave velocity, average about 20,000 ft/sec., has been identified. A series of long holes will be drilled and blasted throughout the block and the wave velocity will be measured.

It is hoped that this exercise will lead to the re-establishment of de-stressing as a tool to combat rockbursts. However, it is only part of a general exercise to control energy. This is:

- Ensure the sequence of mining allows energy to be released at an even rate.
- Make sure that the method of mining and any future methods keeps fill close to the face to restrict closure and thus restrict the total energy release.
- Identify areas of high modulus and soften these so that the dissipation of energy along fracture planes can proceed peacefully.

References

1. G.P. Watson and R. Kerrich. Macassa Mine, Kirkland Lake. The Geology of Gold in Ontario 1983.
2. R.G.K. Morrison. Report on the Rockburst situation in Ontario Mines, Transactions CIM Volume XLV 1942 pp 225 - 272.
3. W.T. Robson. Rockbursts Incidence, Research and Control Measures at Lake Shore Mines Ltd., Transaction CIM Volume XLIX 1946 pp 367 - 376.
4. F. Buckle. The Rockburst Hazard in Wright Hargreaves Mine at Kirkland Lake, Ontario. National Safety Congress - Chicago - 1964.
5. J. Harling. Report on Long Hole De-Stressing in the Kirkland Lake Camp, Private Report.
6. Wilson Blake, Rockburst Mechanics, Quarterly of the Colorado School of Mines - January 1972.

ACKNOWLEDGEMENT

The authors would like to thank Lac Minerals Ltd. for their permission to prepare and publish this paper.

Rockburst control through destressing–a case example

M.P. Board M.S., M.A.I.M.E.
C. Fairhurst B.Sc., Ph.D., M.A.I.M.E.
Department of Civil and Mineral Engineering, University of Minnesota, Minneapolis, U.S.A.

ABSTRACT

A field experiment was conducted in the Star Mine, Coeur d'Alenes, Idaho, to establish the effectiveness of destress blasting to mitigate the severity of rock-bursts. A crown pillar in a cut and fill stope was blasted in order to both reduce the strength and post-peak (unloading) stiffness of the rock prior to extraction. In this way, it was felt, a more stable dissipation of energy released by mining should be possible. The results obtained appear promising.

A brief qualitative discussion of the mechanics of stable and unstable energy changes in rock crushing is presented to introduce the paper.

MECHANICS OF STABLE AND UNSTABLE ROCK FRACTURE IN COMPRESSION

The physical basis for a rockburst has been defined by Cook et. al. (1966) [see Salamon (1973) p. 1081)] as ''the uncontrolled disruption of rock associated with a violent release of energy''. Also the notion of damage to underground workings tends to be implicit in the mining engineers' understanding of the term.

A simple explanation of rock burst phenomena, albeit on a very small scale, is provided by the mechanics of crushing of a rock specimen in a compression testing machine (Figure 1).

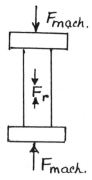

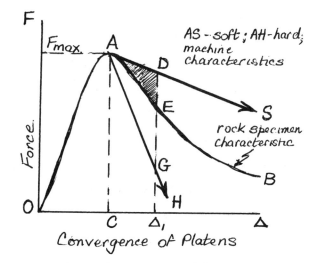

Fig. 1 Brittle Rock Specimen Deformation in Soft and Stiff Testing Machines

Over tne region DA in which the load is increasing, the machine-specimen system is ''unconditionally stable'' [see Salamon 1974, p. 1083] – the applied load is increasing with increasing deformation. Energy is being stored in the machine (i.e., in the frame, pressurizing fluid, hoses, etc.) and in the rock specimen. In the region AB, the system is only ''conditionally stable''. Thus, if AB defines the locus of forces for continued quasi-static deformation (disruption) of the rock specimen, then a machine with an unloading characteristic (i.e., mechanical stiffness as measured by the platen force-platen convergence relationship - a characteristic of the machine that is independent of the rock specimen), such as AS, will release more energy (area under the characteristic AD) for a given deformation Δ_1 beyond the peak value Δ_c then can be absorbed (area under AE) by quasi-static deformation of the rock. The excess energy (area ADE) will be transformed into kinetic energy, i.e., accelerating the rock to violent disintegration. The combination of the soft machine, characteristic AS and the specimen, characteristic AB, results in unstable, i.e., uncontrolled, violent post-peak deformation (disruption). Converseley a 'stiffer' machine (i.e. with less stored energy at a given platen force level), with an unloading characteristic AH, the machine-specimen system will be stable, since energy must be (continuously) added from an external source (increased pump pressure) in order to supply the deficiency between the energy required for quasi-static deformation along AB, and the characteristic AG. The energy deficiency at deformation Δ_1 is represented by the area AEG.

Consider now a system with a rock specimen loaded to L (Figure 2) between <u>rigid</u> platens (i.e. unloading characteristic AC - no energy

stored in the machine). We will assume that a single through-going shear fracture SS_1 develops in the specimen (Figure 2) at L. This behavior tends to appear in specimens of homogeneous, fine grained rocks.

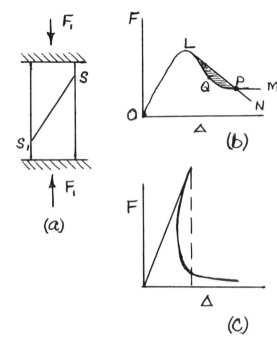

Fig. 2 Deformation of ''Sheared Through'' Specimen

We further assume that i) the normal force-convergence relationship for quasi-static sliding of SS_1 is given by LQM; ii) with <u>no</u> movement of the rigid platens, the strain energy stored in the intact parts of the rock specimen above and below the shear path is released along a characteristic LPN. Here, the specimen alone will tend to disintegrate dynamically due to the excess energy represented by the hashed area LQP, and will tend to deform stably at more advanced stages of deformation.

The specimen can be made to deform stably if the two intact parts of the specimen can be unloaded by reversal of the platen deformation so as to extract the excess energy in this fashion rather than by accelerated sliding on the shear path. The platen force-platen convergence curve for quasi-static deformation of such specimens would then appear as in Figure 2(c). Where the machine is not completely rigid, then obviously a greater reversal would be required in order to abstract the energy stored in the machine in addition to LQP. Load-deformation curves, such as shown in Figure 2(c), have been obtained for brittle rock specimens (eg., granites)(Wawersik 19).

EFFECT OF SPECIMEN SIZE ON STABILITY OF DEFORMATION

It should be noted that the strain energy in the rock is accumulated (stored) in the volume of the specimen, whereas it is dissipated along surfaces (eg., by frictional sliding). Thus if the linear dimension of similarly shaped specimens is doubled, then the volume will increase by a factor of $2^3=8$, whereas the surface area will increase by $2^2=4$. Thus, the energy excess will increase, i.e., the instability will tend to be more violent and catastrophic for larger, initially intact, specimens.

Observation of experiments involving compression of rock specimens indicates that even where the rock is inhomogeneous and multiple fractures develop, deformation tends to become localized along one (or two conjugate) major shear path traversing the specimen. The more inhomogeneous the specimen, the more likely the development of micro-fracturing and small-scale energy dissipation so that, when the final through going shear develops, the more stable it is likely to be, due to the preceding localized dissipation(s) of stored energy.

ROCKBURSTS IN TABULAR EXCAVATIONS

Although violent disintegration of rock is encountered in tunnels, headings and similar ''three-dimensional'' openings the problem of rockbursting in essentially two-dimensional tabular ore bodies is probably of greatest general concern in underground mining.

South African engineers pioneered the application of energy and stability analysis to the practical design of mine layouts and excavation procedures in the burst-prone conditions of Witwatersrand gold-mines. Cook (1966) and Salamon (1974) provide comprehensive accounts of the approaches taken. Excavation of the gold-bearing reef is analyzed essentially as an advancing slit in a homogeneous linearly elastic medium. In such a medium, the equilibrium conditions before and after excavation of a mining panel or, equivalently, the total energy released (by roof-floor closure) is independent of the sequence of excavation, but the increment of energy released at each step of the excavation can vary appreciably. One pattern of extraction, for example, may produce a relatively low energy release in the early stages of excavation, and high releases later, or vice-versa. The optimum pattern is that for which the Energy Release Rate (ERR) (i.e., per unit of extracted reef) is most nearly uniform at all stages of mining. The ERR system has been used to considerable practical benefit in South Africa and other mining regions for almost two decades, with good correlation between rock-burst intensity and calculated ERR.

Computation of the ERR for incremental excavation of a tabular ore-body is relatively simple. The energy released by a linearly elastic rock mass upon excavation of an increment, plan area ΔA is equal to the work which could be extracted from the rock mass by controlling the convergence of the surfaces of the excavation from their original position. Thus, in Figure 3 the mean force F acting on the increment before extraction, and the mean convergence of the increment after extraction provide the end points of the linear deformation (local unloading stiffness). The area under the curve represents the energy released. Several analog and digital procedures have been developed to facilitate computation of the ERR for various excavation patterns. [Salamon (1974), Crouch and Starfield (1983)].

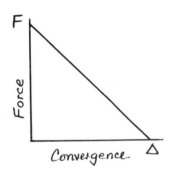

Fig. 3 Energy Released by Increment of Extraction

By analogy with energy changes in the compression test, the line F–Δ in Figure 3 may be taken to correspond to the machine characteristic. Assessment of the overall stability of the energy change due to the

increment of extraction strictly requires a knowledge of the local rock characteristic, i.e., the quasi-static load-deformation behavior for the rock extracted. By relating the rock-burst potential directly to the ERR Cook (1967) tacitly assumes that the rock extracted absorbs none of the energy released from the rock mass, (i.e., the rock characteristic is a vertical drop from peak force to zero force – Figure 4). In the absence of any experimental data on the load-deformation behavior of rock at the (one-sided) face of a mined excavation [this must differ from a (two-sided) pillar characteristic, for which some measurements have been made (Salamon, 1974, p. 1045)], this is perhaps a reasonable, although rough, approximation for a very brittle material such as Witerwatersrand quartzite.

This approximation is not acceptable, however, for a material such as coal. Studies of ''coal-bump'' occurences in U.S. coal mines (Crouch et al., 1973) revealed that ERR values computed on the assumption of a linearly elastic, rigid model (Figure 4(a)) were completely at variance with observed coal-bump locations. Thus, model 4(a) predicts (a) that the rock is elastic at the edge of the excavations and (b) that the highest ERR will occur at the edge of the excavation where stress (and force) concentration are highest.

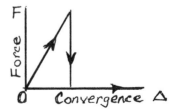

Fig. 4(a) Rock characteristic assumed for Witwatersrand quartzite in calculating ERR

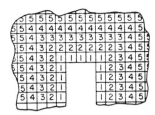

Model of roadway being driven into a block of coal.

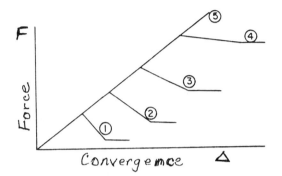

Fig. 4(b) Coal characteristic assumed for coal in calculating ERR

In coal mining, pillar edges and the coal face are frequently observed to 'yield' (i.e. unload) in a stable manner so that stresses are concentrated towards the center of a pillar. For example, Crouch et al., found that it was necessary, in order to obtain reasonable agreement with field data on coal bumps, to divide the coal mass into elements (in plan), to adopt a rock (coal) characteristic for which the maximum load, and unloading slope were functions of the 'degree of confinement' (i.e., distance from a free surface) of each element. This suggests that improved understanding of rock-burst phenomena and their control may require a more careful consideration of the non-linear post-peak load deformation behavior. It is known, from German experience with water infusion in coal bump

prone seams (eg. Sonnenschein) that maintenance of a minimum depth of ''yielded'' coal ahead of a face is a key factor in preventing serious coal bumping and damage at the face.

Understanding of the interction between the ERR (i.e., the energy released by the rock mass) and the energy dissipation function of the rock in the reef or extraction vein is of obvious importance also in the effective use of destressing in ameliorating rock-bursts. It would appear that heavy blasting can introduce fractures into the vein that effectively ''soften'' the rock characteristic, i.e., rendering it more ductile, capable of absorbing a greater proportion of the energy released with less, if any, remaining to produce kinetic effects. The softening effect of blasting would also serve to provoke yielding of the vein at the stope face, developing a ''cushion'' of yielded rock between the face and the highly stressed zone. Successive destress blasts would serve to develop a progressive yielding to maintain the cushion as the face advanced. It is clear that much more needs to be understood concerning the interaction between rock mass energy release and the dissipation of this energy in the rock being mined in order to allow better control of rockbursts.

FIELD DEMONSTRATION

A field application of the concepts discussed earlier was conducted to examine the effectiveness of massive rock destressing as a rockburst control measure. This project, conducted by the U.S. Bureau of Mines in the

95

Coeur d'Alene district over the period 1979–1981, was documented in a recent report[1]. In general, the study consisted of distress blasting of a highly stressed sill pillar, followed by extraction using the undercut and fill stoping method. Rock instrumentation was used along with practical observation, to establish the effects of the blasting.

THE COEUR D'ALENE MINING DISTRICT

The Silver-Lead Mines of the Ceour d'Alene district of northern Idaho have experienced rock bursting problems for nearly 60 years. The narrow (generally 6 meters or less) veins are nearly vertical in dip and are currently mined to depths in excess of 2500 meters. The vein material ranges from a highly fractured sphalerite to an extremely hard, siliceous, argentiferrous galena. The host material consists of interbedded bands of impure quartzite which ranges from a reasonably soft,

fractured argillite to a brittle, highly siliceous, pure quartzite.

The predominant mining method is overhand cut and fill, using either a backstoping or breasting-down technique. Horizontal levels are driven every 61 meters, and the orebody developed by either a lateral and crosscut system, or by drifting along the vein. Mining commences upward from each level toward the above, previously mined stopes in a series of 10 foot cuts. After each cut is completed, it is backfilled, usually with unconsolidated hydraulic mill tailings. In general, a line of three to six stopes (each 61 meters in length) will progress abreast toward the above level (Figure 5). As seen in this figure, a sill or crown pillar is created in this method. When this pillar reaches a height of about 18–20 meters, rockbursting within the pillar and surrounding wall rock will occur, with maximum frequency and magnitude at about 12–15 meters.

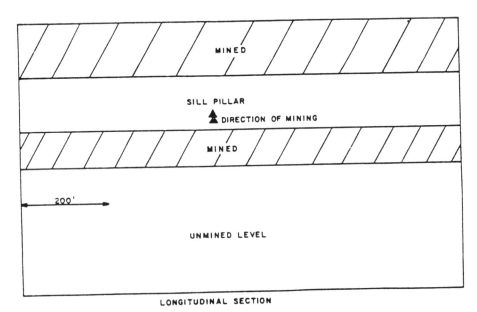

Fig. 5 Geometry For Overhand Simulations
Showing Above Mined Level, and Sill Pillar

These bursts have been recorded at Richter magnitudes of 2 and greater, often displacing 1000 or more tons of rock. The purpose of the present program was to investigate the use of massive pillar destressing techniques in conjunction with undercut and fill mining as a method of controlling pillar bursting.

DESTRESS BLASTING

Destress blasting involves the drilling of blastholes into a pillar or other mine structure, followed by blasting in an attempt to reduce its stiffness and allow a transition from a brittle, bursting failure mode to a non-violent failure characterized by crushing and large displacement. Initial experimentation with destress blastinginvolved drilling of standard 3-5 meterlong jackleg holes with the normal drill round, followed by shooting with the production blast. In most cases the best result which could be obtained was that a burst occured with the blast. Later destressing studies at Hecla's Star Mine were conducted under the sponsorship of the U.S. Bureau of Mines in which an attempt was made at ''massive'' destressing. Here, a section of the orebody was drilled with 76-102 mm blastholes at close spacing, loaded heavily with a gelatin dynamite and shot with millisecond delays. Following the blasts, significant changes in seismic velocity across the orebody were noted, and subsequent mining in this highly burst-prone area was accomplished with little seismicity. The success of this demonstration prompted the present pillar destressing program.

TEST SITE

In September 1980, a line of four backstopes on the Grouse Vein, 6900 level of the Star Mine, had mined to within 15 meters of the 6700 level haulage drift in a more or less flatback geometry (Figure 6). A pillar burst occurred in roughly the center of the pillar, with resulting major damage in the stope as well as the above and below levels. Based upon the history of bursting on this vein, it was determined that additional pillar bursts (with greater magnitude) would occur. These circumstances led the mine management to consider a pillar destressing program prior to completion of stoping. A significant operational problem occurs when mining into blasted ground. The ground is sufficiently broken by the destress blast so that caving can occur, in some instances to the above level. Therefore, it was decided that an undercut and fill technique would be more suitable for use in pillar recovery than the present overhand system. A project was organized with the U.S. Bureau of Mines to provide rock mechanics instrumentation and personnel to study and document this program.

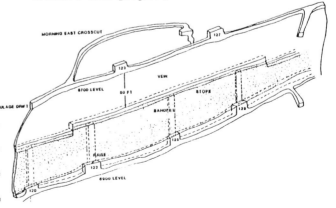

Fig. 6 Isommetric of 6900 Grouse Vein at Time of Sept. 1980 Pillar Burst

THE MINING DEMONSTRATION

Rock Instrumentation

Prior to destress drilling, the rock mass surrounding the pillar was instrumented to monitor rock response to the blast and subsequent mining. This instrumentation included: microseismic monitoring equipment, closure points in the stope and above drifts, fill pressure cells and vibrating wire stress meters in the surrounding rock. A schematic of instrument locations is shown in Figure 7. These instruments were installed and background levels established several weeks in advance of the blasting.

Destress Blasting

The design of the destress blast was the result of a number of compromises. Experience from previous blasts indicated that the heavier the shot, the less the likelihood of future damaging bursts. The mine supervision, however, was concerned (and rightly so) about the practical problems of damage to present workings as well as future problems in mining in heavily broken ground. The resulting compromise is illustrated in Figure 8. A total of 29, 50 to 57 mm holes, (5-8 meters in length) were drilled with jackleg drills on 2-3 meter centers upward into the above pillar. No holes were drilled directly at future raise locations to avoid particularly poor raising conditions. The holes were loaded with ANFO or watergel, depending on hole condition. In either case, a cast primer was first placed in the hole with a 200 grain detonation cord. The explosive was loaded to within 1 meter of the collar. The explosive was detonated using non-electric millisecond delay primers fired from a primacord leg. Poor loading of the holes resulted in only about 400 total pounds of explosive, or about 1 lb per foot of hole.

Results of Destress Blast

The results of the destress blast (even though it was quite small) were immediately evident from the instrumentation as well as from observation. Overnight stope closures ranged

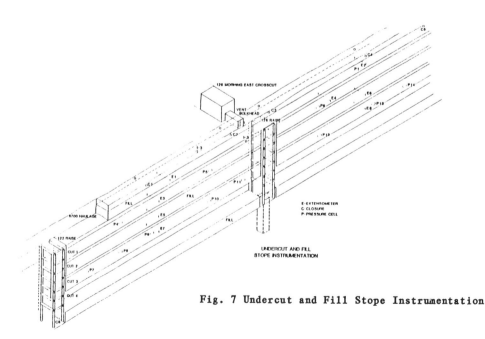

Fig. 7 Undercut and Fill Stope Instrumentation

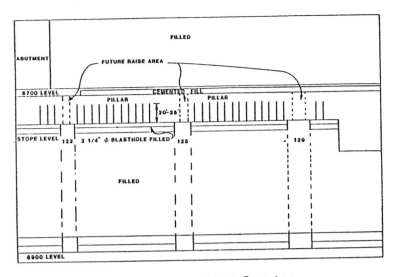

Fig. 8 Destress Blast Geometry

from 8 to 38 mm. Horizontal pressure in the cemented fill in the above drift increased by approximately .9 MPa. Immediately following the blast, little microseismic activity was recorded, however, the geophone array was widely spaced, and thus only activity with significant energy level would be sensed. In the days and weeks following, the increase in closure rate, and decrease in pillar stress meters was dramatic (Figures 9 and 10). From an observational standpoint, the decrease in pillar stiffness was obvious. Prior to start-up of mining (and after the blast), significant squeeze on stope timber could be seen. Headings were crushing, and caps were cracked. From all observations, the pillar behavior had been altered from brittle-elastic to plastic.

Following the blast, raises were driven to the above level at each chute. Normally, this operation would be very hazardous from the bursting standpoint, but, only a bad back was encountered. Extraction of the pillar then continued using the undercut and fill technique with cemented sandfill. Numerous mining procedural problems was encountered due to poorly consolidated fill, however, there were no severe ground problems. As seen in Figure E, as the pillar height was reduced below 10 meters, the closure rate was extremely high, resulting in as much as 460 mm of closure over this final 12 meters of mining. The stress meters (Figure F) continued to show a decrease in stress, eventually coming to equilibrium with a 6 meter pillar remaining.

The seismic activity was minimal during pillar recovery. Numerous low magnitude, low frequency events were recorded, and were attributed to frictional sliding along fracure surfaces in the crushing pillar. Only one small burst was recorded, some 30 meters into the wall rock, resulting in minor sloughing of rock from stope walls.

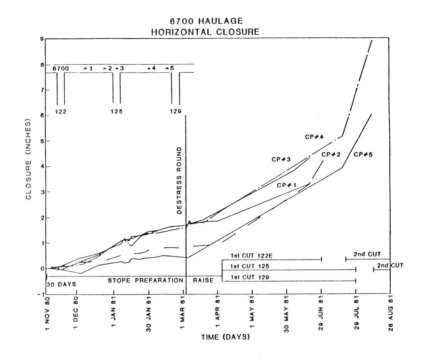

Fig. 9 Closure of the 6700 Haulage Drift. Time
Period from Preparation to Mining the First Two
Undercut Levels.

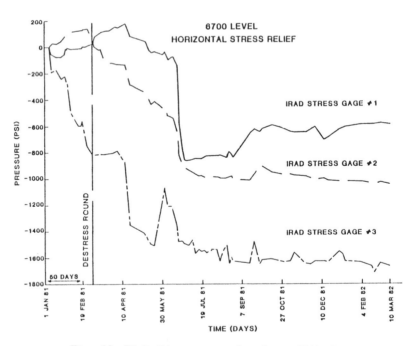

Fig. 10 IRAD Stressmeter Results, 6700 Level
Gage 1, 2 and 3

CONCLUSIONS

Both instrumentation and practical experience indicated that the pillar destressing initiated a non-violent failure of the sill pillar. The use of undercut and fill mining enabled recovery of this pillar, even though it was highly crushed during the final cuts. The details of how the destress results in a decrease in pillar stiffness is unknown. Since the fairly widely spaced blastholes provide no free face for breakage, we must assume that the creation of new fracture surfaces is not important. The most reasonable explanation would appear to be separation or dislodging of fracture surfaces from their original mating positions, allowing sliding at lower stress levels. Further studies with more extensive instrumentation are required to provide further information on this phenomenon.

ACKNOWLEDGMENTS

We wish to acknowledge the work of W. Blake who was instrumental in the daily activities of the field demonstration discussed here. His previous work, along with that of J.T. Langstaff of Hecla Mining Company provided the basis for the fieldwork. The funding for the field work was provided by the U.S. Bureau of Mines, Spokane Research Center under the direction of Mr. M. Jenkins.

References

1. D.D. Bush, W. Blake and M.P. Board (1982) Evaluation and Demonstration of Underhand Stoping to Control Rock Bursts, Terra Tek Contract Report #H029-2013.

2. Cook N.G.W. (1966), The Design of Underground Excavations, Proc. 8th Symp. on Rock Mechancis, Univ. of Minnesota. Ed., C. Fairhurst, pp. 167-193, AIME New York, 1967.

3. Crouch S.L. and Fairhurst C. (1974), The Mechanics of Coal Mine Bumps and the Interaction between Coal Pillars, Mine Roof and Floor", Final Report, Research Contract H0101778, Feb. 1973, U.S. Bureau of Mines, pp. 25-26.

4. Salamon M.D.G. (1974), Rock Mechanics of Underground Excavations. General report Theme 4, Proc. Third Intl. Cong. on Rock Mechanics, Vol. 1B, Int. Soc. Rock Mech., Denver. Published by U.S. Nat. Acad. Sci., Washington, D.C. (1974), pp. 951-1099.

5. Crouch, S.L. and Starfield, A.M. (1983), Boundary element methods in solid mechanics, George Allen and Unwin (1983), pp. 322.

Rockburst phenomena in British coal mines

N.J. Kusznir B.Sc., Ph.D., F.G.S.
University of Keele, Keele, Staffordshire, England
I.W. Farmer B.Sc., Ph.D., C.Eng., F.I.M.M., F.I.Min.E., M.I.C.E.
University of Newcastle upon Tyne, Newcastle upon Tyne, England

SYNOPSIS

Rockburst phenomena occurring in British coal mines are summarised, and two phenomena and their associated seismicity are described in greater detail:

(a) pillar collapses associated with long wall coal mining at depth in the North Staffordshire coalfield

(b) gas and coal outbursts in anthracite seams in the South Wales coalfield.

The dissimilar mechanisms are discussed and proposals made respectively for design and prediction.

INTRODUCTION

Rockbursts occur when rocks in the vicinity of mine excavations fail rapidly during stress changes associated with the mining operation. Such failure, accompanied by sudden release of stored strain energy, is a brittle fracture phenomenom and is usually associated with high stresses, low confinement and strong rocks. This is not a combination immediately associated with British coal mining conditions. However, over a period of time, certain phenomena which recognisably fit the definition of rockbursts have been experienced in British coal mines. Some of these have constituted a significant hazard to safety and production.

There are several forms in which rockburst phenomena in Britain can manifest themselves, and although they may have different causes, they may usefully be summarised:

(a) rock or coal failure in the working area, usually at a face or roadway side and resulting in the violent expulsion of rock from the face, often accompanied by floor heave and emissions of gas, have been described at depths of 730m in the South Staffordshire coalfield (South Staffordshire and Warwickshire Institute of Mining Engineers, 1933). These were attributed to excessive stress on the exposed abutments.

(b) pillar collapse between workings has been
described at Barnborough Colliery (Phillips 1944)
and Hickleton Main (Shepherd and Kellet, 1973) in
the Yorkshire coalfield. Both occurred in the
Parkgate seam below the thick strong Parkgate
Sandstone, and although uncommon in Britain
probably have most in common with typical hard
rock rockbursts.

(c) pillar or abutment collapse, remote from the
active face due to superposition of stresses from
the mining operation on residual stresses from
previous workings. This has occurred in deep
mines at 1000m in the North Staffordshire coalfield (Kusznir et al 1980a,b). Seismicity has
been felt at the surface although not at underground workings protected by overlying wastes.
Similar events have been described by Mashkour
(1976) in the Midlothian Coalfield.

(d) rotational collapse of cantilevered beds of
strong rock, forming the roof of longwall faces,
about the face abutment line. This is a common
occurrence in caving long wall operations where
sandstones or other strong rocks occur in the
roof strata sequence. Case histories are
described by Lampl (1983) in this symposium.

(e) activation of old fault planes due to mining
activities adjacent to, and particularly under-
neath major cross measures faults. Examples of
this have been found in the North Staffordshire
coalfield (Al-Saigh, 1981). Such failure tends
to generate seismicity but does not cause serious
mining problems.

(f) outbursts of gas and coal which occur in
South Wales anthracite mines (see for instance
Pescod, 1947). It may be argued (Farmer and
Pooley 1967) that these events fall outside the

classical definition of a rockburst. Nevertheless
the level of strain energy released and the fact
that a substantial quantity of material is
ejected from the face sometimes suggests a
similar mechanism.

Of these phenomena, those which have recently
caused the greatest problems are the outburst
phenomena of South Wales, and those which have
created the greatest interest are the pillar
failures of North Staffordshire. The latter,
whilst causing few mining problems has resulted
in considerable seismic activity at the surface.
These two types of failure will be discussed in
greater detail.

PILLAR FAILURE IN NORTH STAFFORDSHIRE

In the North Staffordshire coalfield, underlying
the densely populated area of Stoke-on-Trent,
numerous coal seams have been extracted at depths
greater than 1000m from a deep basin structure.
Extraction has been accompanied by seismic
activity at the surface, which in the years
1975-1977 and 1980-1981 reached magnitudes of the
order of 3.5 and Mercalli Intensities of between
V and V1. The main problem was an environmental
one, associated with seismic activity, and the
investigation (Kusznir et al, 1980a, 1982) was
originally aimed at identification of the source
of the seismic events rather than investigations
of rockburst mechanisms.

A surface seismic network (Figure 1) was
used to investigate the location of the activity
which was centred around rapidly advancing long
wall workings in the Ten Feet Seam at Hem Heath
Colliery. These workings, situated at a depth of
about 1000m, were overlain and underlain by
extensive workings in other seams. In particular,

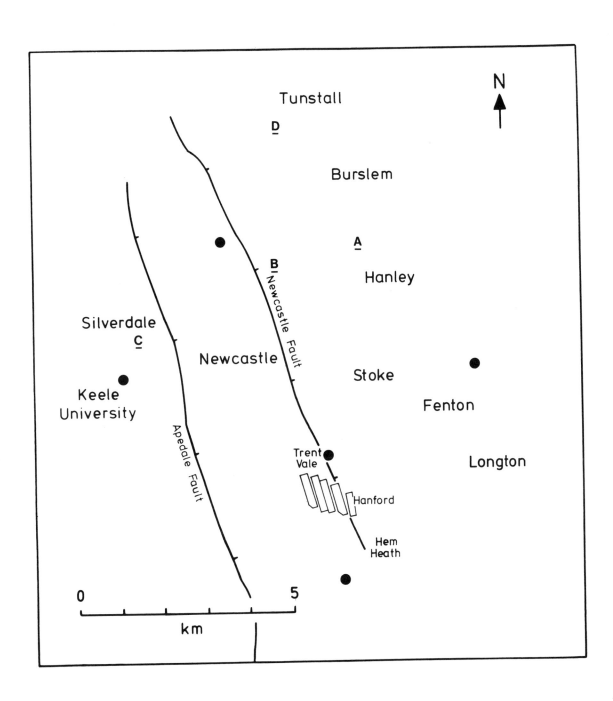

Figure 1   A map of the Stoke-on-Trent conurbation showing location of long wall panels in the Trent Vale/Hanford area mined from Hem Heath Colliery and other seismic areas A-D. Major faults and seismometer networks are also shown.

pillars in the Moss Seam 200m above, and the Bowling Alley Seam 20m below, were crossed by a series of long wall faces in the Ten Feet Seam.

Seismic activity was monitored continuously during the mining of panels 205 and 206 (Figure 2) and tremor hypocentres and magnitudes determined from the recorded data. Tremor hypocentre locations occurring during the mining of both panels are shown on Figure 2 in plan and in North-South and East-West section. It is evident that tremors lie adjacent to active mine workings at depths between 600 and 1000m.

An additional higher resolution small aperture network was operated for a short period of time during the mining of panel 206. Tremor hypocentres determined using this network during one month of mining activity are shown in Figure 3. First motion analysis was used to show that two different source mechanisms existed; one of a shear type and the other implosional. These different mechanisms are distinguished in Figure 3.

Data analysis together with considerations of face position and the location of the workings in other seams led to the following conclusions (Kusznir et al 1980b):

(a) Tremors were associated with coal extraction in the Ten Feet Seam.

(b) Tremors did not lie on any of the major faults within the mined area.

(c) Tremor hypocentres moved with the advancing coal face.

(d) Of 709 tremors detected instrumentally during the working of panels 205 and 206 only 8% were felt at the surface. None were felt on the face.

(e) Felt tremors occurred as the Ten Feet faces passed over or under pillars in the Bowling Alley and Moss Seams.

(f) The greatest tremor activity - consisting of many small events - occurred as the Ten Feet faces passed over a pillar in the underlying Bowling Alley Seam.

(g) A critical span in the direction of face advance of 250m appears to be necessary for the onset of seismicity.

(h) Tremor hypocentres appear to be displaced to the East of the active Ten Feet panel by about 200m.

(i) Events with larger magnitude have shear type source mechanism and coincide with those felt at the surface.

(j) The smaller, unfelt, events have an implosional source mechanism.

It is possible to explain the seismicity observed in terms of the basic mechanics of rock collapse during long wall mining and interseam interaction. The smaller events with an implosional source mechanism are almost certainly generated by waste collapse. These do not constitute any underground hazard and are the inevitable and normal consequence of waste caving.

The larger magnitude events having a shear source mechanism are generated by pillar failure in adjacent seams. The largest shear events were situated at the edge of the 200m wide Moss Seam pillars, and may be explained by shear failure of the partly confined and residually stressed pillar edges, subjected to the abutment stresses from the Ten Feet Seam workings. Collapse of the narrower pillar in the Bowling Alley seam generated smaller shear events. This may be attributed to the likely earlier partial failure of the pillar under residual mining stresses

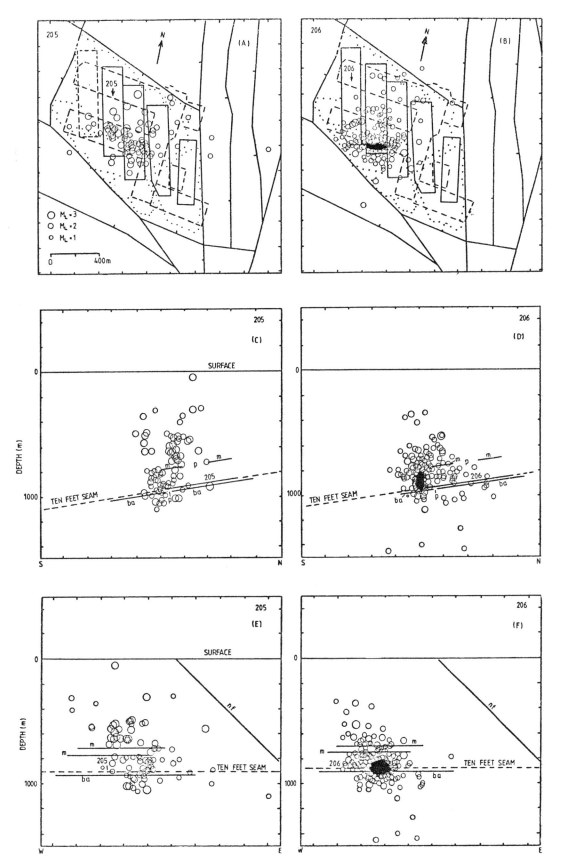

Figure 2   Tremor hypocentres occurring during the mining of panels 205 and 206 in the Ten Feet Seam from Hem Heath Colliery.

(ab) Plan View
(cd) Vertical section perpendicular to face
(ef) Vertical section parallel to the face
Both panels advanced to the south. Previous workings in the Moss (m) and Bowling Alley (ba) seams are shown together with the location of faults.

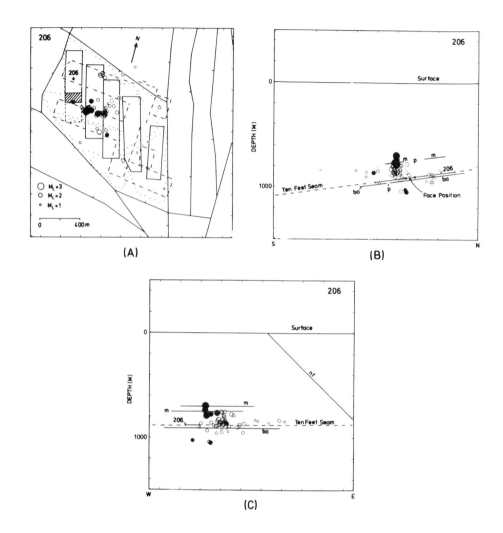

Figure 3   Higher accuracy tremor locations occurring during one months mining of panel 206.

(a) Plan View.
(b) Vertical section perpendicular to face.
(c) Vertical section parallel to face.

Solid circles represent events with a shear source mechanism while open circles represent events with an imlosional mechanism. The face position is represented by hatched ornament.

(see Kusznir et al 1982 for a more detailed discussion).

The occurrence of these quite strong events associated with interaction between active and barrier pillars is probably a feature of deep multi-seam mining. Because this is rare in Britain it has been thoroughly researched. Although it does not appear to constitute a serious problem at the moment, it is a factor which should be considered in the design and layout of deeper workings. In particular the methods which are used to design barrier pillars to protect underground roadways (Wilson and Ashwin 1972) virtually ensure the likely existence of residual pillar stresses and create the conditions for future pillar collapse if interaction of deep workings occurs.

There have been other seismic events in North Staffordshire associated with active long wall mining. Some events are summarised in Figure 4 for the two year period 1975-1977. The seismicity in areas A, C and D mainly associated with workings from Wolstanton Colliery has a similar shear source mechanism to the events observed at Hem Heath Colliery, and has a similar pillar collapse origin. The seismicity in area B is of interest in that it appears to be generated by slip movement of a pre-existing fault plane, induced by long wall mining under the fault.

It is worth remarking that the seismicity in area A, generated magnitudes of the order of 3.5 later in 1980 and caused considerable environmental concern. One of these events was also felt at the face.

OUTBURSTS IN SOUTH WALES

Outbursts in the South Wales coalfield - and in other coalfields throughout the world - constitute a serious operational problem in high rank coals of high vitrinite content. They also occur in some evaporite deposits, particularly potash. In the South Wales coalfield the anthracite coals in the Gwendraeth Valley have been affected by outbursts. Today the problem is confined to Cynheidre Colliery where outbursts occur with a frequency of one every few weeks.

An outburst is a rapid relase of pulverised coal and gas into a mine roadway or face. Outbursts often occur at the ends of faces or in headings although they can also occur in mid-face. They are preceded by considerable noise followed by a burst of the coal face-wall which releases large quantities of gas and coal, often from quite a small opening or fracture. The amount of coal released - virtually as a fluidised bed propelled by desorbed gas, often at a rapid rate - can be several hundred tonnes.

There are several geological and tectonic features which are invariably associated with outbursts. These include:

1. The strata have been tectonically disturbed often by a thrust fault or fold

2. The tectonic disturbance has been accompanied by shearing of the coal which changes its state to a finely sheared mass

3. Significant amounts of adsorbed gases are contained on the surfaces of the sheared coal particles.

4. The roof and floor rocks and the coal abutment on all sides of the sheared zone are strong and intact

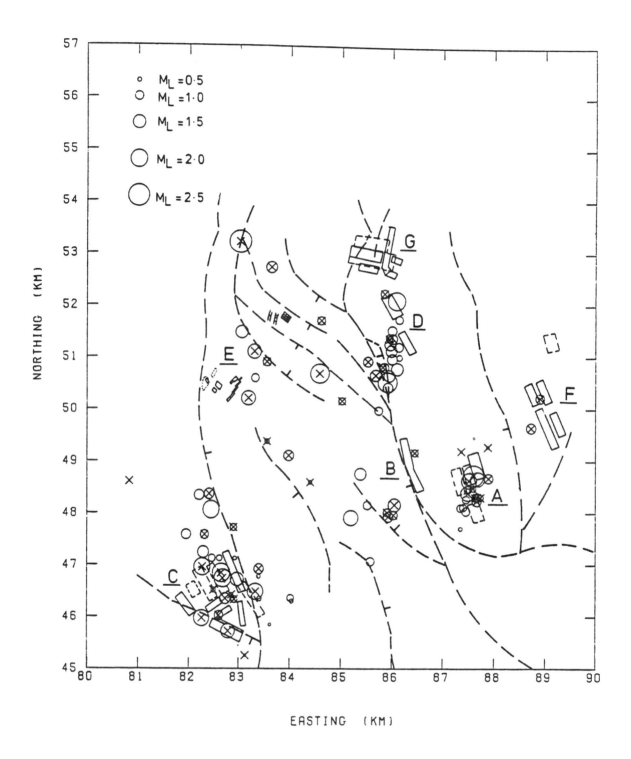

Figure 4   The location of earth tremor epicentres occurring during 1975-1977 in the northern
            part of the North Staffordshire coalfield. Rectangles correspond to active long
            wall coal panels. Seismicity in areas A - D is associated with the extraction of
            coal. (from Al-Saigh, 1981)

It can be argued that the reason why outbursts tend to occur less often in the centre of long wall faces is that high abutment stresses leading to vertical fracturing of the coal face allows the gases to desorb gradually. In more confined circumstances the conditions may be created for quite rapid fracture of coal abutments under some redistributed stresses, but mainly from stresses induced in the strong coal by the high pressures from gas starting to desorb as the face coal dilates slightly.

An example of the breakdown characteristics of anthracite from outburst zones at Cynheidre is given in Figure 5. These tests were carried out on 150mm long by 75mm diameter specimens in the manner described by Farmer (1983). These illustrate some of the unique characteristics of the material. At low confining pressures it has low strength, failing in tension on cleat planes with quite large dilation. The brittle characteristics of the material are not over emphasised. At higher confining pressures, however, the tension failure is inhibited by confinement and shear failure at a high axial stress magnitude and in noticeably brittle manner occurs. The potential for explosive energy release under these conditions is evident.

The onset of the outburst process should therefore be marked by considerable activity in the restraining coal abutment. Scholz (1968) among others has shown that this will be accompanied by seismic events. If this is the case then seismic monitoring of outburst areas should form a basis for outburst detection and control.

A surface network of geophones has been installed at Cynheidre - using similar equipment to that in North Staffordshire - in order to investigate seismic activity associated with outbursts. Results have not yet been analysed in full, but observations are expected to confirm similar observations published by Styles (1983). These are

(a) Precursive seismic activity occurs prior to the occurrence of outbursts.

(b) This activity is of microseismic magnitude.

(c) The precursive activity is probably in addition to that associated with general long wall mining.

(d) It has not yet been possible to identify the source mechanism of the seismic activity.

(e) The seismic events occur with increasing frequency prior to the outburst. This precursive activity suggests that the monitoring of the microseismic activity could provide a useful predictive tool for outbursts. For this it would be necessary to define a threshold seismic activity level associated with the onset of an outburst.

(f) Outbursts may occur spontaneously or may be induced by precautionary firing. Only spontaneous events show precursive seismic activity.

The fact that precursive seismic activity occurs before the spontaneous events suggests that the outburst is initiated or evolves through brittle fracture of a rock material. Typical curves of accumulated microseismic activity during breakdown of a brittle material have been obtained by Scholz (1968) and others (see Figure 6). It is likely that the activity envelope for anthracite will follow this pattern. It is significant that Leighton (1982) in a speculative approach to out-

111

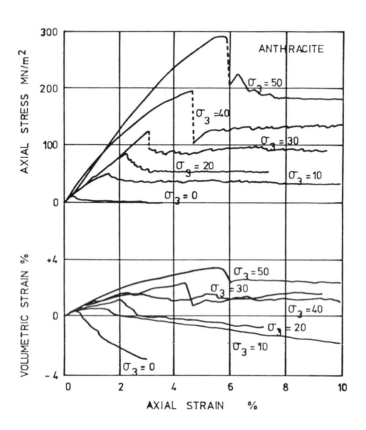

Figure 5   Axial stress-strain and axial stress - volumetric strain curves at confining
pressures of 0-50 MN/m² obtained by Godden (1982) from specimens of anthracite
from typical outburst areas.   Specimen size was 150mm by 75mm diameter and rate
of testing 2 x 10⁻⁵sec⁻¹.

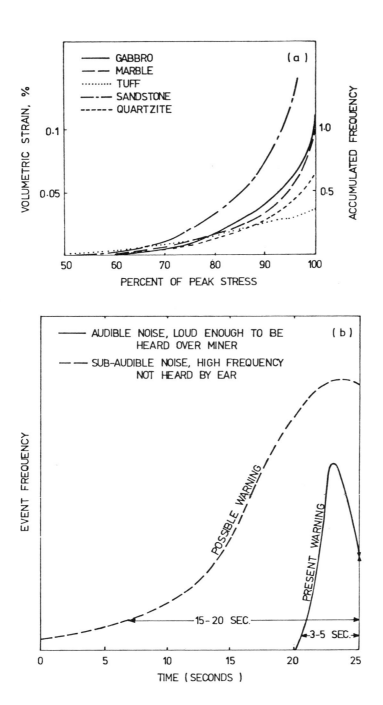

Figure 6 (a) Plots of dilation (the difference between measured volumetric strain and computed 'elastic' volumetric strain) against percentage of fracture stress and a dimensionless representation of accumulated frequency of microseismic events for five rocks (after Scholz, 1968). Note that Scholz was only able to separate dilation and accumulated frequency at stress levels very close to failure.

(b) A hypothetical relation between event frequency and time before an outburst (after Leighton, 1982).

113

burst initiation and prediction proposes a similar relation between rock noise rate and time.

If the precursive seismic activity is associated with rock dilatancy and the growth of microcracks it is tempting to speculate on the role of the desorbing gas on the stability and growth dynamics of the microcracks. The gas if desorbed at pressure into the cracks may accelerate the crack growth process so precipitating the explosive disintegration of the rock and the violent release of gas and coal fragments.

CONCLUSIONS

(a) Rockbursts do occur in British coal mines but are not a major problem.

(b) In-seam pillar failures, typical of hard-rock mining rockbursts have been reported at depths greater than 700m in areas where strong thick sandstone beds overlie the mine workings.

(c) The most common type of rockburst occurs at depths of the order of 1000m where pillars in abandoned workings contain residual stresses from previous workings. Superposition of additional abutment stresses from over - or under-mining can lead to pillar collapse, in areas where strong sandstone beds are not present. This is likely to become a problem with deeper coal mining but can be relatively easily alleviated by correct design.

(d) Outbursts in South Wales are a diminishing problem - with diminishing mining activity. They are a stress related phenomenon, associated with sheared strata and propagated by desorbed gas. They are preceded by increasing microseismic activity and can be monitored and predicted through this activity.

(e) Monitoring of seismic activity associated with rockbursts and outbursts is a useful investigative and predictive tool.

References

AL-SAIGH, N.H. 1981. Mining induced seismicity of the North Staffordshire Coalfield, unpublished Ph.D. Thesis, University of Keele.

FARMER, I.W. 1983. Engineering behaviour of rocks 2nd Edn. Chapman & Hall, London.

FARMER, I.W. and POOLEY, F.D. 1967. A hypothesis to explain the occurrence of outbursts of coal, based on a study of West Wales outburst coal. International Journal Rock Mechanics Mining Science, Vol. 4, pp. 189-193.

GODDEN, S.J. (1982). Bursts and outbursts in coal mines unpublished MSc Thesis, University of Newcastle upon Tyne.

KUSZNIR, N.J., ASHWIN, D.P. and BRADLEY, A.G., 1980a. Mining induced seismicity in the North Staffordshire coalfield, England. International Journal Rock Mechanics Mining Science and Geomechanics. Abst., 17, pp. 45-55.

KUSZNIR, N.J., FARMER, I.W., ASHWIN, D.P., BRADLEY, A.G. and AL-SAIGH, N.H., 1980b. Observations and mechanics of seismicity with with coal mining in North Staffordshire, England, Proceedings 21st U.S. Rock Mechanics Symposium, pp. 163-171.

KUSZNIR, N.J., AL-SAIGH, N.H. and FARMER, I.W., 1982. Induced seismicity resulting from roof caving and pillar failure in long wall mining, in Strata Mechanics, Developments in Geotechnical Engineering, Vol. 32, edited by I.W. Farmer, Elsevier, 1982, pp. 7-12.

LEIGHTON, F., 1982. The search for a method to provide warning of coal and gas outbursts, Proceedings 2nd Conference Ground Control, Morgantown, West Virginia University, pp.117-123.

MASHKOUR, M., 1976. A seismological study of a mining area, Ph.D. Thesis, University of Strathclyde.

PHILLIPS, D.W., 1944. Rock bursts or bumps in coal mines. Transactions of Institution of Mining Engineers, 104, pp. 55-84.

SHEPHERD, R. and KELLET, W.H., 1973. Strata behaviour, a study of faces and roadways in workings under a sandstone roof liable to rock bumps. Colliery Guardian, 221 (3), pp. 93-102.

SOUTH STAFFORDSHIRE AND WARWICKSHIRE INSTITUTE OF MINING ENGINEERS, 1933. The occurrence of bumps in the thick coal seam of South Staffordshire, Transactions Institution of Mining Engineers, 85, pp. 116-147.

STYLES, P. 1983. Microseismic precursors to a spontaneous outburst in Cynheidre Colliery, Dyfed, South Wales (Abstract), Geophysics J.R. Astr. Soc., 73, 299.

WILSON, A.H. and ASHWIN, D.P. (1972). Research
into the determination of pillar size. <u>Mining</u>
<u>Engineer</u>, Vol. 131, pp. 409-430.

115

Longwall planning with reference to rockbursts

B.N. Whittaker B.Sc., Ph.D., C.Eng., F.I.M.M., F.I.Min.E.
Department of Mining Engineering, University of Nottingham, Nottingham, England

SYNOPSIS

Rock bursts in coal mining operations have been encountered in many countries. In some cases the degree of geological disturbance is the predominant factor although in others mining depth can equally play a major role. The geological history of the coal deposits and the ensuing degree of disturbance both geologically and from subsequent mining have major influences. The review indicates the importance of a comprehensive geological assessment coupled with an indepth assessment of the influence of a chosen mining method. The mining method plays a major role in those cases where rock burst situations have arisen and special adaptations for improved control are related to. An appreciation of the stress configurations resulting from mining operations is also shown to be of major importance in planning coal mining layouts in geological formations especially to rock burst phenomena.

Specific case histories are reviewed in order to demonstrate particular principles concerned with planning mining layouts particularly in longwall operations. The case histories reviewed are taken from Europe and North America although the type of geological situations are frequently encountered in other parts of the world.

INTRODUCTION

Rock bursts are not unique to hard rock mines. Although the frequency and magnitude of rock bursts create special problems in some hard rock mining situations, notably in South Africa and India remarkable progress has been achieved in gaining a better understanding of their mechanisms and conditions favouring their occurrence. In coal mining situations, mining depth has always been very much less than in hard rock underground mines when rock burst problems have occurred. The Ostrava-Radnavice coal mining region in Czechoslovakia is currently experiencing special rock burst problems at depths below surface of 500-600 m; the coal seams in that coal field have been subjected to substantial geological disturbances. Rock bursts have been encountered in underground coal mines over many years in several countries, although their occurrence has been on a fairly minor scale when considering the full extent of underground working of coal. The geological setting of a particular coal deposit has a major influence on

whether rock bursts may occur but the mining method also plays a significant role and must be taken into account. Mine planning in such situations requires certain factors to be carefully considered and these are discussed in the paper.

NATURE OF ROCK BURST PROBLEMS IN COAL MINES

The effects of rock bursts in coal mines can be localised or even regional depending upon several factors. The most frequently encountered hazard is that of a sudden outburst or movement of coal into the mining excavation, and in many cases this is accompanied by release of a significant volume of gas and dust. The physical characteristics of coal, especially its brittle nature coupled with low strength to resist fracture and tendency towards violence during failure are important features. However, some coal deposits when subjected to major geological disturbances such as over-thrusts, significant and extensive shearing and intensive folding, can develop a high degree of fracture with low shear resistance and be highly charged with gas and dust particles within impervious interfaces created by geological shearing movements. Such hazardous pockets if tapped by a mining excavation can give rise to outbursts of gas and fine coal.

The presence of major thicknesses of strong competent rock immediately overlying a coal seam can create rock burst situations which usually exhibit sudden failure and bursting of the coal into the mining room, roadway or working face. The induced stress field associated with the mining operations requires careful consideration. In addition, a type of failure which can fall into the category of a rock burst is that of sudden and large scale failure of the immediate roof behind a working longwall face since the effects of the resulting shock wave can trigger support instability and other problems.

MAIN FACTORS INFLUENCING OCCURRENCE OF ROCK BURSTS

The main factors can be categorised under two broad headings, namely geological and mining induced stresses.

Geological factors

The extent and degree of geological disturbance

experienced by a coal deposit can create the necessary conditions for rock/coal bursts and in many circumstances these are frequently accompanied with sudden releases of gas and fine coal dust. Substantial shearing can give highly polished slip surfaces and create significant quantities of fractured material possessing low inter locking resistance which under certain conditions can be mobilised to flow into mining excavations; the accompanying gas released during such a process promote the flow characteristics of the sheared material. Such flow processes are sudden owing to the highly stressed condition of the previously fractured (and in some cases pulverized material).

Extensive folding of coal seams can also give rise to the conditions which create rock/coal bursts associated with the working of such seams. Some anthracite seams in South Wales and France are well-known for their proneness to outbursts.

Major geological intrusions have been known to cause high stress anomalies to occur and give rise to localised rock burst conditions.

Some coal seams appear to be more prone to rock bursts and coal outbursts than others; the relative permeability of the immediate roof and floor of the seam play an important role since for outburst conditions to significantly develop parts of the disturbed seam need to act as a closed gas reservoir.

The thickness of the coal seam is not a principal factor since thin seams (1 m and less) have exhibited these problems, although in a given geologically disturbed mining field there is probably a greater risk in thicker seams providing the necessary conditions discussed above are present.

Earthquake effects have been known to trigger rock bursts in coal mines and the most recent example of this type of occurrence has been in China.

Mining induced stress factor
The state of mining induced stresses within a given mine configuration can result in the necessary conditions favouring rock bursts. High stress abutments associated with ribsides and pillars can give rise to hazardous zones which can cause rock bursts even when working a subsequent seam within the proximity of the previously worked seam. Special problems have been encountered with rock burst situations in the Ruhr Coalfield, West Germany especially where a significant thickness of competent sandstone has overlain the coal seam being currently developed; the high state of stress within the vertical proximity of an abutment zone has caused the coal sides to burst suddenly into the mine roadway. This has been encountered more frequently at depths exceeding 800 m and although strata stress abutment zones due to previous mining operations have been the major contributory cause a common factor has been the presence of strong sandstones overlying the new seam.

The occurrence of rock bursts in this category is most common where a development heading is driven into a highly stressed abutment zone; strain relief is frequently the triggering mechanism with sudden bursting of the coal sides or

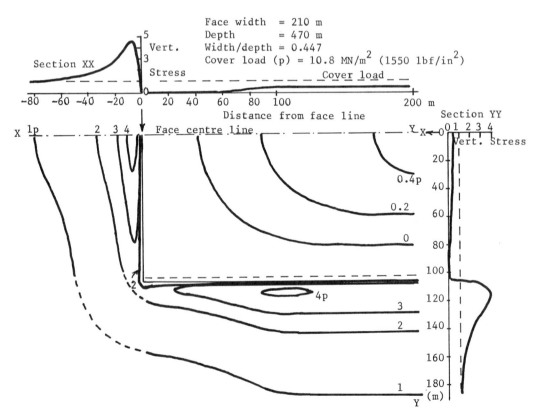

Figure 1. Distribution of vertical stress produced by longwall extraction showing front and flank abutment zones[1].
Note – Data based on average depth and longwall dimensions for East Midlands Coalfield, U.K., and field observations.

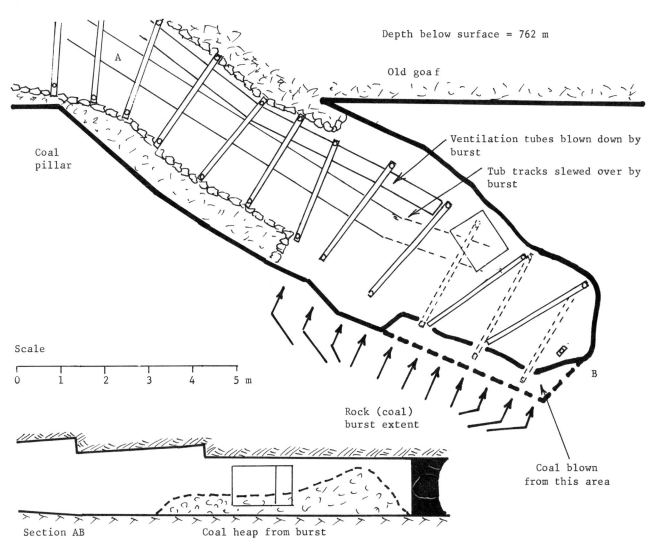

Depth below surface = 762 m

Old goaf

Coal pillar

A

Ventilation tubes blown down by burst

Tub tracks slewed over by burst

Scale

0   1   2   3   4   5 m

B

Rock (coal) burst extent

Coal blown from this area

Section AB   Coal heap from burst

Figure 2. Denaby Main Colliery (Parkgate Seam) rock burst incident.

it may take the form of sudden floor upheaval. Certain types of mining layout can give rise to increased proneness for rock bursts to occur.

CASE HISTORY FROM DENABY MAIN COLLIERY
Denaby Main coal burst[2]
A coal burst occurred in a development heading in the Parkgate Seam at Denaby Main Colliery, U.K. on 9th January 1957 at a depth below surface of 762 m. Figure 2 illustrates the main features of the site of the burst. The roof consisted of very strong sandstone (25-30 m thickness). The coal burst was described as occurring very suddenly and one side of the heading appeared to blast away almost filling the heading cross-section at one point. The driven dimensions of the rectangular profile were 2.7 x 1.45 m and the support system consisted of props and bars. The rock burst was confined to the coal seam; the roof and floor were unaffected. Witnesses to the incident described the burst as taking the form of a heavy 'bump' which is a general term applied to this type of phenomenon in coal mining operations. An air blast accompanied this coal burst. The workmen driving the heading prior to the burst described the coal in the locality of the rock burst as being very workable and readily 'crumbling' (less than one hour before the incid-

ent). Figure 1 (section AB) shows the extent of coal released by the burst which at its highest point was about 25 cm from the heading roof.

General comments
The strain energy within the coal and immediate sandstone roof was the prime cause of the coal burst. The coal burst zone was located 5-8 m from the old goaf ribside edge. The flank strata pressure abutment zone was a major contributory factor to the creation of the highly stressed coal whose strain energy gave rise to the burst. Undoubtedly, the sandstone roof also plays a role in such incidents, particularly in allowing strain energy to be built up and give the potential hazard of sudden release under certain circumstances.

This type of rock burst problem is encountered in the Ruhr Coalfield, Western Germany and in the Donbass Coalfield, U.S.S.R. Special efforts at the Skochinsky Mining Research Institute, Moscow have been directed towards using micro-seismic emissions from the coal as a means of detecting the possible occurrence of a coal burst. The research findings indicated that there was a clear build-up in micro-seismic activity just prior to the burst taking place.

119

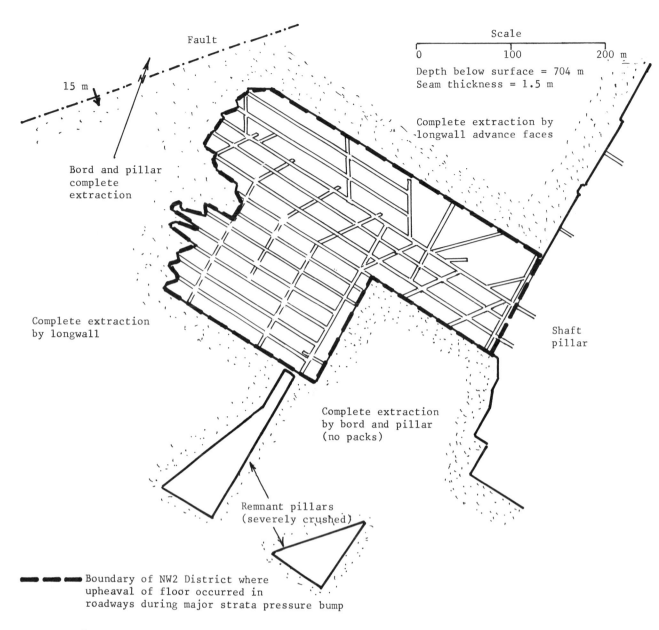

Figure 3. Barnborough Main Colliery (Parkgate Seam) strata pressure bump incident.

CASE HISTORY FROM BARNBOROUGH MAIN COLLIERY
The type of rock burst incident which occurred at Barnborough Main Colliery, 24th April 1942 was concerned with sudden upheaval of the floor in the roadways developed in a spur of the Parkgate Seam off the main shaft pillar[3]. The depth of the seam was 703 m whilst the thickness was 1.5 m; the immediate roof section consisted of about 30 m of strong competent sandstone. The affected area of the mine is shown in Figure 3. It had overall dimensions of the order of 300 x 200 m although by virtue of its shape the narrowest part was about 100 m wide where it adjoined the shaft pillar. The pillared area was in the process of systematic extraction back towards the shaft.

The rock burst incident took the form of sudden upheaval of the floor with most of the roadways becoming completely impassable; there was no noticeable warning from the surrounding rocks with failure of the floor occurring very quickly (there was no time to take evasive action). In addition to the floor upheaval, there was evidence of some pillar-edge failure also although on a

much lesser scale than the floor. However, if the floor had been very strong and competent, then it is more than likely that the pillar edges forming the roadway sides would have failed causing complete (or almost complete) sudden closure of the roadways. The degree of closure of the roadways was different over the affected area of NW2 District; a few roadway sections were relatively lightly affected especially near to the pillar extraction operations. Roadways well within the pillared complex were worst affected with complete closure and most being impassable. Access for rescue and recovery operations within this area needed to be along the pillar edge (within the seam) adjacent to the severely distorted roadways. The remnant pillars to the south of NW2 District were severely crushed and indicate the extent of the incident.

The cause of the upheaval was the sudden loading of the pillared area by overlying strata; the strong sandstone beds overlying the seam will have tended to span over wide expanses of the previously extracted workings on three sides of the NW2 District. Progressive pillar extraction to the west of the district probably triggered large scale failure of a significant expanse of overlying strata; the 15 m fault further to the west will also have played a role in providing a relief line to permit strata loading to be transmitted to the NW2 District. The fact that this pillared area was a spur off the main shaft pillar made the NW2 District vulnerable to high strata loading pressures.

The use of pillar extraction by this method immediately ceased at this mine. Longwall advance mining employing a more systematic extraction pattern was adopted.

This form of rock burst has been encountered in many coal mining countries especially where the immediate strata overlying the coal seam consists of extensive thicknesses of strong sandstones (or similar rocks). Remnant pillars have posed special problems, both within the current seam and other seams above or below. Irregularly shaped remnants create most problems, especially the acute corner regions.

SUDDEN LARGE SCALE PILLAR FAILURE

The failure, on a large scale, of coal pillars whose strength has been exceeded by the strata load transmitted to them forms a special category of rock burst in respect of coal mining operations. The problem has been mainly encountered in room and pillar operations, either

(i) where the size of the pillars has not been adequate for long-term stability taking into account progressive development of the mine,

or

(ii) where the overlying strata are very strong with extensive spanning properties and their continuity are disrupted by geological faulting or even natural failure due to subsidence, then sudden loading of pillars in parts of the mine can result in failure by bursting.

Employing a recognised pillar design formula with an adequate factor of safety against failure is of paramount importance, but the geological setting of the mine must be also carefully considered. Major pillar failures are very sudden and are frequently accompanied by an air blast hazard. Pillar extraction operations associated with room and pillar mining can give rise to irregular strata loading of coal pillars skirting the extraction area.

The rib pillars employed between longwall coal faces are not so vulnerable to bursting owing to their high degree of constraint along their major axis.

PRESSURE AND SHOCK BUMPS IN COAL MINING OPERATIONS

The terms 'pressure bump' and 'shock bump' have been used in coal mining operations to identify the type of failure by coal bursting.

Pressure bumps

This refers to failure of a coal pillar, mining room or roadway by overloading. In the case of mine pillars, as employed in room and pillar layouts, should the strata loading exceed the overall strength of the pillars, then pillar failure could be sudden and violent but it depends upon the strength of the immediate floor

and roof. A strong roof and floor could encourage violent failure if the coal pillars were overloaded but a very weak roof and floor would encourage localised yield and give some constraining effects with failure clearly displaying some creep properties.

Shock bumps

This refers to sudden loading of a coal pillar, or group of pillars, by failure of large expanses of overlying beds causing sudden transfer of strata loading to the coal pillars. The problem mainly occurs where large uncaved areas immediately adjoin pillared areas and where the overlying rocks are very strong.

The above terms are of a general nature and particular mining situations need to be judged in relation to several factors when considering potential rock burst hazards in coal mining operations.

LONGWALL PLANNING CONSIDERATIONS WITH REFERENCE TO ROCK BURSTS

The major considerations to be taken into account when planning longwall operations in situations judged to be prone to rock (coal) burst risks are listed.

(i) The type of geology especially the structural features of faulting and folding, and degree of geological disturbance are major factors. Equally important is the type of coal, since the higher the rank the greater the susceptibility to bursting under high states of stress. The nature of the geological stresses perhaps play the greatest role in influencing coal burst situations especially outbursts of fine coal and gas. Such mining situations demand a careful appraisal of the geological aspects and especially the historical evens which could lead to such bursts occurring. Some of the high risk severely folded and contorted parts of a coal seam should be approached only with caution, and in some cases avoided.

(ii) Strong sandstone beds overlying a currently working coal seam present the first stage of risk in connection with rock (coal) bursts in coal mining operations. Its mode of control following the mining operations needs to be carefully assessed.

(iii) Pillared areas on a large scale should be avoided, since rooms and roadways within such an area could be at risk if extensive caving is carried out in the immediate vicinity of the pillars.

(iv) Rib pillars are the most stable form of support and offer a high measure of protection against pressure and shock bump risks. The roadways skirting the rib pillar (i.e. gate roadways) possess a high measure of protection against closure by sudden failure owing to the strain relief around the roadway. However, roadways within the rib pillar could be at risk, since these are the most vulnerable to sudden closure during such a bump situation.

Design formulae[4] for determining the stress

levels within coal rib pillars are given as equations (1) and (2).

$$\bar{\sigma} = \frac{9.81\gamma}{1000.p^2}\left\{\left[(p+w).D - \tfrac{1}{4}w^2.\cot\phi\right]p\right\} \quad \ldots (1)$$

for $w/D < 2\tan\phi$ and

$$\bar{\sigma} = \frac{9.81\gamma}{1000.p^2}\left\{p\left[p.D + D^2\tan\phi\right]\right\} \ldots (2)$$

for $w/D > 2\tan\phi$

where

γ = average density of the overburden
ϕ = angle of shear of roof strata at edge of longwall extraction and measured to vertical (taken as 31°)
p = width of rib pillar
w = width of longwall extraction
D = depth below surface.

A p/D ratio of at least 0.1 to 0.15 should be considered for stability and protection against rock burst risk.

(v) When developing part of the mine where there might be a rock burst risk as mining progresses, the main roadways should be located in stress relieved ground such as skirting a pillar with pack support on one side, or even in a totally extracted area (which offers the maximum protection). Roadways in solid ground would be the most vulnerable to such a hazard.

(vi) Longwall advancing is the most favoured in in such circumstances in view of the forward relief it gives and greatest portection against sudden closure to roadways. Advance location of pockets of high stress and highly charged gas pressure reservoirs presenting a potential outburst hazard is very difficult however since borehole probing is still the most positive means of detecting them.

(vii) Standard longwall retreat mining layouts should be avoided when there is a major risk of rock burst since the development headings will be within the seam and generally only lightly supported. Props and bars offer minimal resistance to sudden side wall closure. Longwall retreat headings located over remnants and rib edges of workings in other seams represent risk situations. The 'Z' system of longwall extraction offers scope for longwall retreat but without the inherent risk which development headings have in solid ground where rock burst hazards exist.

(viii) Planned extraction should be systematic avoiding remnants and the working of vulnerable spurs avoided if at all possible. The use of rib pillars between successive longwall advance faces offers the greatest protection against the rock burst risk.

(xi) Due consideration should be taken when driving headings into high strata pressure abutment zones in coal seams having a potential bursting hazard; props and bars offer the least support resistance against side-wall failure. The support system within the most vulnerable zone (5-20 m from the solid ribside) should be intensified to give adequate protection against sudden side wall failure. The use of arch steel sets or rock bolts with mesh should be employed in such situations.

CONCLUSIONS

Some coal mining operations experience rock burst problems as do some hard rock mines. In the case of coal mines, the properties of the coal and the fact that gas and dust can accompany the bursts makes the problem more complex. Longwall mining is the most effective method of reducing the rock burst risk in coal mines and providing the working layout is planned for systematic extraction with adequate protection from suitably designed rib pillars, then a safe and efficient means of extraction can be achieved. The geological situation needs to be carefully assessed together with the implications of the proposed mine plan in relation to stress fields and the degree of protection afforded by rib pillars and de-stressed areas to access roadways and current working faces.

ACKNOWLEDGEMENT
The author expresses thanks to various mining engineers within the National Coal Board for many useful discussions in relation to rock burst problems in coal mines. Special thanks are recorded for valuable help given by Mr. S. Lewis, Senior Strata Control Engineer, N.C.B. Headquarters Mining Department.

REFERENCES
1. Whittaker, B.N. Evaluation of the design requirements and performance of gate roadways. The Mining Engineer, February 1979, pp 535-548.
2. Ditchfield, R. Report on Coal Burst in D4/B5 Heading at Denaby Main Colliery. N.C.B. Report, North Eastern Division, No. 3, Area, January 1957.
3. Humphrys, H.J. Report on causes of and circumstances attending the upheaval of floor which occurred on 24th April 1942 at Barnborough Main Colliery, South Yorkshire. HMSO Cmd 6414, 1943, 10 pages.
4. Whittaker, B.N. A review of progress with longwall mine design and layout. Proceedings of Society of Mining Engineers Conference on "State of the Art of Ground Control in Longwall Mining and Mining Subsidence" pp 77-84. A.I.M.M. and P.E. Publication, Editors Chugh and Karmis.

Effect and limitations of abnormal loads on roof supports

F. Lampl C.Eng., F.I.Mech.E.
Dowty Group Mining Division and Dowty Mining Equipment, Ashchurch, Tewkesbury, Gloucestershire, England

CONTENTS: Introduction
Experiences
Investigations
Discussion Of Results
Recommendations

INTRODUCTION

This paper relates the experiences of a support manufacturer when longwall mining in conditions where "rockbursts" or "weighting beyond the rated capacity of the supports" have occurred. It distinguishes between these two conditions and describes the work carried out to find a solution to increase the acceptance of the supports to such conditions. It also points out the design limitations and the necessity for special mining measures to reduce the severity of such conditions.

EXPERIENCES

The occurrence of rockbursts in the South African gold mines has been known for many years. Work by the South African Chamber of Mines has resulted in a better understanding of the causes and in recommendations for equipment and mining methods to reduce the severity of rockbursts and the damage resulting from them. Over the last few years longwall coal mining has also suffered such incidents, often resulting in considerable damage to the roof supports and personal risk to the operators. Prediction of these conditions is poor. Even after the event it is not always clear whether damage has been the result of impact or excessive roof weight released over the supports. Case studies no doubt will improve our understanding but more fundamental work is desirable to enable us to predict abnormal caving conditions prior to planning and installing a longwall. The following is a short account of occurrences with which the writer was personally involved.

1. No. 4 Mine, Tatung Coal Field, China

Face data:- No. 2 Seam
Face length of 150m supported by 4 x 550 ton chock shields
Extracted height 3 metres
Retreat working

Geological data:-
No. 2 Seam is overlain by 4.4m of grey sandstone and gravel and thereafter by 41.7m of white and grey sandstone.

In normal operations the first 7m of strata above the seam caves readily.

The next 11/12m of roof strata caves at irregular intervals; "beams" of up to 50m have frequently occurred.

Events:-

Rapid closure and forward movement of supports on several occasions resulting in considerable damage to the legs and advancing rams of the supports.

FIG. 1 TYPICAL DAMAGE TO LEG IN CHINA

The first incident was reported after 159m advance from the start of the face. Further incidents of varying severity occured at irregular intervals. From questioning of the mine personnel it appeared that both shock loading and heavy weighting was the cause of these events.

Similar damage was reported and seen at Mine 12 (in the same coal field as Mine 4). In this case no impact loading was reported but sudden and silent closure of the supports resulted in severely damaged legs.

2.　Texasgulf, Wyoming, U.S.A.

This mine is extracting Trona, sodium sesquicarbonate from a depth of 400 metres. The immediate strata above the Trona is dolomitic marlestone. The usual method of extraction is by room and pillar method using continuous miners.

Towards the end of 1982 they installed a 60m shortwall face. A continuous miner cuts an 11 ft web and the face is supported by 4 x 600 ton chock shields. These shields stand 5½ ft from a reference rail and advance immediately behind the continuous miner. An extendible cantilever supports the remaining 5½ ft.

Events:-

Caving is extremely heavy.

Enormous blocks can be seen in the goaf, one was estimated to weigh 3,000 tons. On several occasions groups of supports were projected towards the face. Rapid lowering of supports was also reported. Damage to advancing rams and their attachments resulted. There was also damage to some legs. In the opinion of the mine personnel the causes of these events are:-

a)　Roof breaking over area of goaf shield. Impact on sloping surface projects shield forward.

Separation of large roof blocks normally causes a drop in the roof stress above the supports. The large roof blocks impacted the sloping surface of the goaf shield before the support had time to return to setting pressure thereby projecting the shield forward.

b)　Roof breaking over canopy. No impact but roof load beyond support capacity resulting in rapid closure of supports.

c)　Heavy blocks being dislodged in the goaf and rolling into the rear of the shield.

3.　Very recently a severe "rockburst" was reported to have occurred at Ceska Armada Colliery, Czechoslovakia.

In this case rockburst conditions had been predicted and the 4 x 500 T chock shields had been equipped with rapid yield valves. In an earlier incident sudden closure of the supports had been reported but no ill effect to the equipment on the face. On the 27th April, 1983 another incident occurred with disastrous effects. Supports, conveyor and machine were severely damaged for 60 metres of the face and 340 metres of the roadway closed and several lives were lost. It is reported that it was the floor which had heaved up suddenly.

INVESTIGATIONS AT DOWTY

Normal yield valves used in powered roof supports are designed to allow a moderate closure rate of the support. Should geological conditions demand rapid closure of the support, this could result in permanent damage to the legs and structure of the support. At Dowty Mining Equipment Limited, we have carried out a full investigation to seek methods of providing protection for supports when subjected to these sudden high loads.

From our investigations, we have found that there are two types of condition which can cause rapid closure of the supports:

Case 1   Sudden impact loads which have the effect of hammer blows, due to heavy caving or due to sudden stress changes in the strata.

Case 2   Roof loads experienced by supports in excess of their support yield load, causing the supports to lower faster than the normal yield valve can cope with.

It had been practice to provide "rapid yield valves" when mining in conditions where "rockbursts" are expected. This improved the ability of the supports to lower faster under excessive yield conditions but it was doubtful whether it would provide sufficient protection to impact loading. This led us to carry out investigations to study the effects and provide remedies for the cases 1 and 2 above.

Part 1 - Rapid Closure Due To Impact Loads

To simulate high impact loads, we obtained the assistance of the Central Electricity Generating Board (C.E.G.B.), which has a test rig where heavy weights can be dropped from varying heights on to the legs to be tested. All the legs tested were of the same specification, and details can be found in Appendix 1.

Procedure:-

All the power set legs to be tested were assembled in a protective box, and the complete assembly placed inside the vertical test rig. A floating top cap was bolted loosely to the leg, and this loose fitting enabled the load to self-align on impact with the 5 tonne weight.

FIG.2   TEST SITE

FIGURE 2
The C.E.G.B. Test Site
The rig consists of a steel tower and is capable of dropping a 5 tonne weight from varying heights of up to 10 metres from the point of impact.

A pressure transducer was built into the leg and connected to instrumentation which recorded a time pressure curve. All the legs were coupled to a release yield module set to relieve at 535 bar with a length of hydraulic hose. All legs were pressurised to a setting pressure of 320 bar prior to each test. The input signals from the pressure transducer were recorded on a four channel tape recorder, and replayed via a filter unit to a recorder. The information was then transferred to an X -Y plotter.

Altogether, 47 tests were carried out on the test rig, with variations of the following parameters:
a)   Drop height
b)   Leg extensions
c)   Type of yield valve.

FIG.3 LEG IN PROTECTIVE ENCLOSURE

FIG.4 LEG READY FOR TEST

Results:-   FIGURE 5

The tests shows that even when the weight was dropped from a moderate height the pressure recorded by the transducer inside the leg was in excess of $1\frac{1}{2}$ times the yield pressure (> 800 bar). When the drop height was increased, failure eventually occurred through dilation of the outer tube, allowing the piston seals to extrude, thereby losing the pressure inside.

During testing, however, we noticed that the pressure rises very much less when the legs are set to near their full extension, and that maximum pressures are recorded when the legs are extended by a small amount.

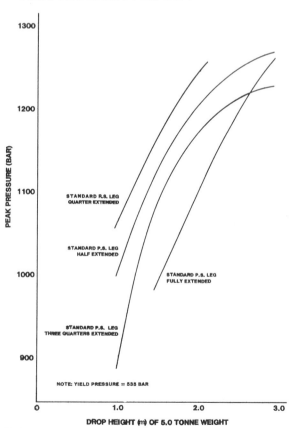

FIG.5

RESULTS OF IMPACT TESTS ON STANDARD LEGS

126

FIG.6
DAMAGE TO LEG UNDER EXTREME HIGH IMPACT

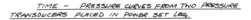
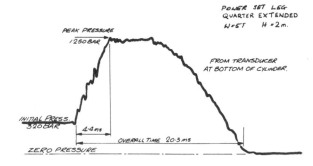
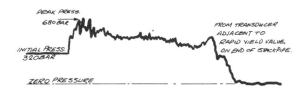

FIG.7 PRESSURE RECORDINGS
(RAPID YIELD VALVE FITTED TO STACK PIPE)

FIGURE 6 - Shows the failure of a leg extended by a small amount and subjected to a high impact loading.

FIGURE 7- Shows the pressure recordings taken inside the leg and outside the leg next to a rapid yield valve. It demonstrates clearly that under impact conditions yield valves offer only little protection to the legs.

The results also demonstrated that, under identical test conditions the increase in pressure in a fully extended leg was less than that in a partially extended leg. This led us to increase the volume of fluid inside the leg by drilling through the piston of the leg, thus flooding the inner tube. The leg now holds a permanent column of fluid and its effect becomes more significant as the power set leg closes. (See Figure 8)

Tests were then carried out with legs having flooded inners and the results plotted and compared with those of standard legs.

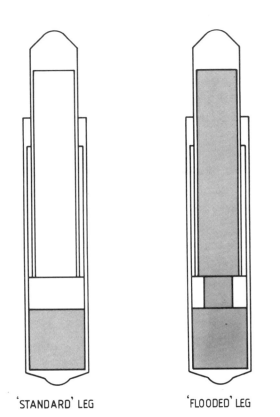

'STANDARD' LEG   'FLOODED' LEG

FIG.8 FLOODED INNER TUBE

TABLE 1

	Volume of Fluid Beneath Piston (Litres)	Permanent Volume of Fluid Stored in Inner Tube (Litres)	Percentage Volume Increase
Fully Extended	39	36	192%
Half Extended	19.5	36	285%
Quarter Extended	9.7	36	471%

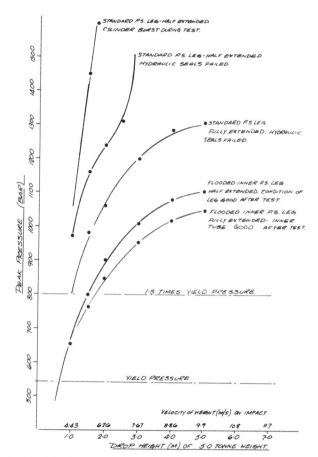

FIG. 9 COMPARISON OF STANDARD LEGS WITH LEGS HAVING FLOODED INNERS

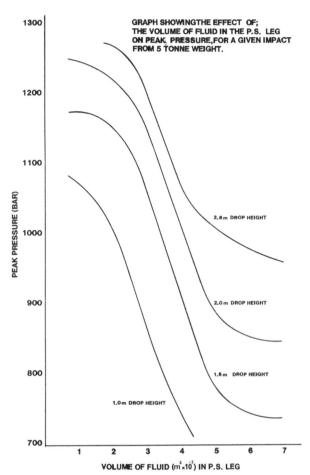

FIG. 10 SUMMARY OF TEST RESULTS

FIGURE 9 and FIGURE 10 - Show this comparison and clearly demonstrates the benefit of flooded inner legs in surviving a much wider range of impact loads.

FIGURE 11 - Is a Time-Pressure Graph comparing the results of a standard leg with those having a flooded inner. The reduction in peak pressure is approx. 40%.

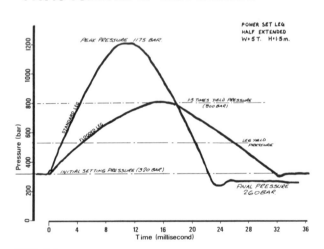

FIG. 11
TIME/PRESSURE GRAPH OF TYPICAL TEST

Part 2 - Rapid Closure Due to Excessive Roof Loads

Load experienced by powered roof supports in excess of the rated yield load causes the supports to lower. If this load is excessive and un-impeded, then the supports will lower faster than can be permitted by the normal yield valves and damage will result to the power set legs as the pressure build-up is unrelieved. Where

mining is under-taken in such conditions, the supports must be designed to meet the excessive roof loads, and this must be such that the energy imparted to the support by the excessive roof load is dissipated quickly enough to allow the load to reach a stable equilibrium before damage occurs to the powered support.

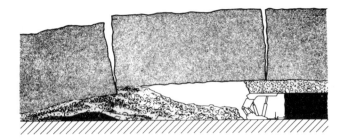

FIG. 12 DIFFICULT CONDITIONS UNDER A SANDSTONE ROOF

FIGURE 12 - Shows a possible condition when mining under a very strong strata.

The supports may be at a near yield load when the roof breaks over the supports. The result is a long block of massive proportions resting in the goaf at one end, and on the support at the other end. The support is forced to yield rapidly until an equilibrium condition occurs, when the slab has fallen sufficiently to become further supported by material in the goaf.

The speed of the support closure is critical as the flow through the valve must be limited to prevent pressure build-up inside the leg to more than $1\frac{1}{2}$ times yield load. This critical velocity depends on the magnitude of the load and the discharge rate of the yield valve. It is therefore essential that 'rapid yield valves' are fitted to powered roof support legs when mining under these conditions.

Our investigation into flow rates of valves were carried out in a purpose-designed rapid closure rig (Figure 13). The tests carried out involved the following:

a) Normal yield valve

b) Rapid yield valve fitted externally to the leg

c) Rapid yield valve fitted internally to the leg.

The results of the above test are shown in Figures 14 and 15.

FIG. 13 RAPID CLOSURE TEST RIG

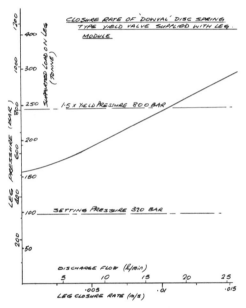

FIG. 14
CHARACTERISTIC OF NORMAL YIELD VALVE

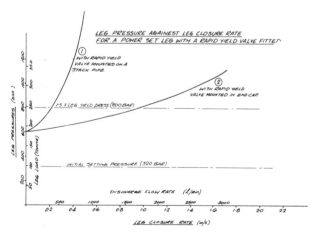

FIG. 15 CHARECTERISTIC OF RAPID YIELD VALVE

It can be seen that the maximum closure rate of a normal yield valve is approximately 0.01 metres per second, which corresponds to 17 litres per minute for a 195mm bore leg if we limit the critical pressure to 1½ times yield load. From the characteristics of the rapid yield valve, the closure rate achieved by the valve in the top cap is 4½ times that obtained with the valve fitted to the stack pipe. The deterioration of flow is largely due to the high pressure drop across the stack pipe and the change in direction of flow. The maximum closure rate with the valve fitted in the top cap is now 1.0 metres per second, i.e. a flow rate of 1750 litres per minute for a 195mm bore leg. This represents an increase in performance of 100 times when compared with the conventional type yield valve.

It may be more practical to consider the distance a support is able to close under rapid yield conditions before reaching critical pressure inside the legs.

The following work is an approximation and does not take into account the compressibility of the floor, the fluid inside the legs and elasticity of the support structure.

Let W = Weight of broken strata above support

F = Resistance to support closure at any instant

X = Distance travelled from rest

Then force = mass x acceleration

$$(W-F) = \frac{W}{g} \times \frac{d^2X}{dt^2}$$

but $F = F_1 + KV^2$

(assuming square law for approximation)

where

F = Support resistance at nominal yield

K = Characteristic valve constant determined experimentally

V = Instantaneous velocity of closure

Investigation and rearrangement gives

$$X = \frac{W}{2gk} \quad \log e \quad \frac{W - F_1}{W - F_1 - KV^2}$$

Figure 16 shows the amount the leg can close when subjected to loads above its rated capacity before it reaches a critical pressure of 1½ times the yield load. The graph assumes that the legs are at yield when the break in the roof occurs.

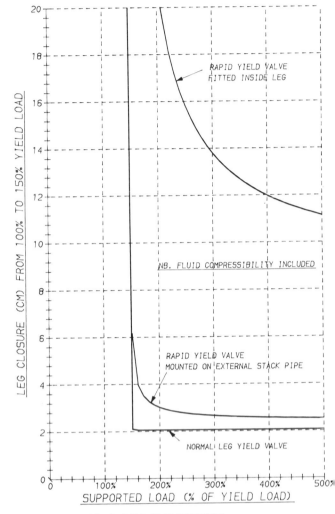

FIG. 16 CRITICAL LEG MOVEMENT

DISCUSSION OF RESULTS

The work described in the foregoing does indicate the protective measures which should be incorporated in a support that is intended to work in abnormally heavy strata conditions.

In the case of impact loads we have seen that rapid yield valves offer little protection but the provision of an accumulator in the form of additional fluid in the legs greatly reduces the pressure rise and the resulting damage to powered support legs. The use of gas accumulators for impact conditions could even be of greater benefit.

Rapid yield valves are essential where roof loads substantially in excess of the support capacity are expected to be encountered. They allow a faster closure rate of the support. With rapid yield valves we are gaining distance which the roof can descend before it finds a solid support in the goaf.

We have also been made aware of the limitations of such protective measures.
How big is the impact?
What is the magnitude of the excessive roof load? If these are beyond the limitations of the support design, other measures which change the character of the strata must be introduced.

In the case of No. 4 Mine, Tatung, China, as described in the beginning, the mine management resorted to water infusion of the sandstone ahead of the face. This considerably weakened the sandstone and reduced the distance of the roof in cantilever behind the supports from 47 - 57m to 13 - 17m. This, together with roof drilling at both face ends and the protective measures introduced on the supports allowed longwall mining to continue.

In the case of Texasgulf, the provision of rapid yield valves, setting loads at 90% of yield, guaranteed set and a protective build up on the goaf shield and exceptional housekeeping and face discipline, have reduced the incidents of shields being moved forward to acceptable limits. Mining measures to reduce the heavy blocking of the roof are also being investigated.

RECOMMENDATIONS

Where heavy caving conditions or even "rockbursts" are anticipated ALL protective measures should be provided as it is likely that the roof supports will be subjected to both types of loading. The supports should have the following features:

1. Support of adequate capacity to ensure roof breaks behind supports.

2. Setting load as close as possible to yield load.

3. Guaranteed set.

4. Large volume power set legs with accumulator effect.

5. Rapid yield valve of the highest practical flow capacity. They should be additional to normal yield valves and set 10% above the nominal yield load.

6. 4 leg shield configuration preferred for better stability.

7. Projection of goaf shield beyond the break-off line to be as little as possible.

8. Goaf shield to be reinforced.

I hope that this paper will also help planners of longwall faces to understand the limitations of the protective measures that can be provided in roof supports against abnormal roof weightings and take this into account when deciding on the location and method of mining of future longwall faces.

APPENDIX 1 - POWER SET LEG SPECIFICATION

Type

Single telescopic,
double acting

Yield load

162 tonne

Length

3165mm maximum
1862mm minimum

Outer Tube

Bore

195mm

Wall thickness

17.5mm

Maximum column of fluid
beneath piston

1303mm

Material

Weldable steel tube
0.2/0.28% carbon, 1.2/15%
manganese, min. U.T.S. 660
N/mm^2, silicon killed steel
supplied by British Steel
Corporation

Inner Tube

Bore

185mm

Wall thickness

12.5mm

Fluid column (if flooded)

1530mm

Diameter of piston bore
(if flooded)

90mm

Material

Weldable steel tube 0.2/0.28%
carbon, 1.2/1.5% manganese, min.
U.T.S. 750 N/mm^2, aluminium
killed steel supplied by British
Steel Corporation.

Rockburst problems in Norwegian highway tunnels—recent case histories

A.M. Myrvang Dr.ing.
Mining Division, The Norwegian Institute of Technology, Trondheim, Norway
E. Grimstad Cand.real
The Norwegian Road Authority, Oslo, Norway

SYNOPSIS

The Norwegian fjord countryside necessitates a large number of highway tunnels. Many of these tunnels have encountered serious rockburst problems. Traditionally, tunnels have been constructed along and around the fjords. This has often created rockburst problems due to gravitational "valley" side stresses. Today, long tunnels are constructed through the mountain ranges between fjords. This has in many cases created rockburst problems due to horizontal, geological stresses. This paper describes case histories presenting both types of rockburst problems. In all tunnels rock stress measurements have been performed at a relatively early stage of tunnelling, and rockburst "forecasts" for the rest of the tunnel have been made. The types, effect, cost and longtime reliability of rock bolting and other stabilization methods are discussed.

INTRODUCTION

The rugged Norwegian fjord countryside necessitates a large number of road tunnels. During the last 100 years several hundred tunnels have been driven. The length of the tunnels may vary from a few tens of meters to more than 7 km. The tunnels are mainly in hardrock, and traditionally the great majority of the tunnels are unlined, except from occational zones of weakness that may have needed lining of some kind. However, during recent years several tunnel projects have encountered serious rock burst problems due to high stresses over major parts of the tunnels' length.

STRESS CONDITIONS

The rock burst problems are in general caused by two types of virgin stress conditions:

1) Traditionally, many tunnels have been driven along and around fjords or valleys. Often the mountain side is very steep with a height of 1000-1500 m. This will create gravitational "valley side" stresses as indicated on fig. 1:

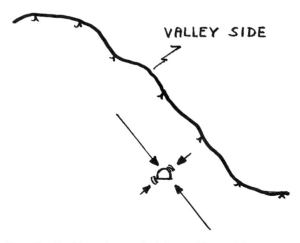

Fig. 1 Rockburst created by valley side stresses. A tunnel is usually driven quite close to the surface. The gravitational stress field here is very anisotropic. This will create large stress concentrations as indicated on the figure, resulting in rock bursts or spalling due to shear failure.

2) During recent years it has become common to construct tunnels through the mountain range between fjords or valleys instead of going around. Fig. 2 shows the situations.

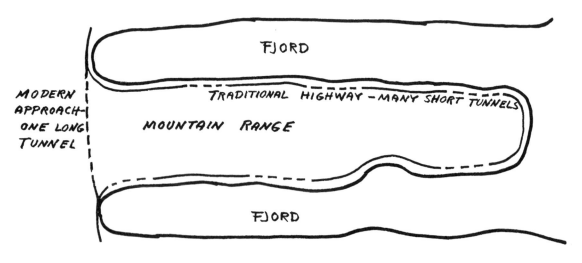

Fig. 2 Traditional practice versus modern approach

Most of the tunnels are situated in Western Norway in precambrian gneissic rocks. In most cases the overburden is quite high, and when the first tunnel of this kind was driven, rock bursts problems were expected in the sidewall due to high vertical stress. However, it turned out that rock burst or spalling occured in the roof of the tunnel at a very moderate overburden. This indicates high horizontal stresses perpendicular to the tunnel axis. This was verified by triaxial in situ measurements. Similar measurements later on in three other tunnels in the same province show that the major principal stress σ_1 is approximately horizontal and may have a value in order 30-40 MPa. The orientation is parallell to the mountain range, while the intermediate and minimum principal stresses are representing the "valley side" stresses as shown on fig. 3.

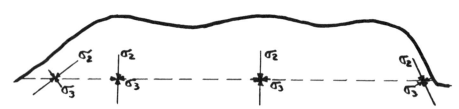

Fig. 3 Intermediate and minimum principal stress orientation

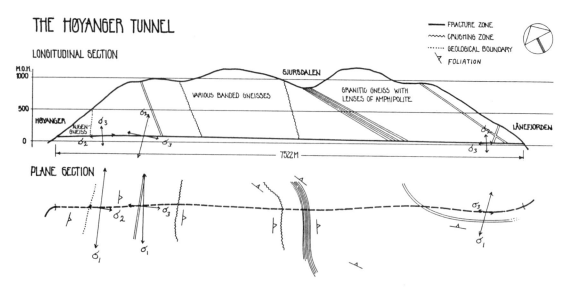

Fig. 4 Plan and longitudinal section of the Høyanger-Lånefjord tunnel

134

The horizontal stress seems more or less constant over the length of the tunnel (except from the near surface zones influenced by erosion and local topography).

CASE HISTORIES
The Høyanger-Lånefjord tunnel

The excavation of this tunnel started in Nov. 76 at the Høyanger side, and in Nov. 77 at the Lånefjord side, and it was put into operation in 1982. The total length is 7522 m included avalanche cover and portals, and the tunnel is the longest highway tunnel in Scandinavia. Fig. 4 shows plan and longitudinal section of the tunnel.

The rock mass consists mainly of different types of gneiss. Lenses of amphibolite occur frequently, and more seldom a few layers of quartsite.

The uniaxial compressive strength of rocks is in the range 60-200 MPa dependent on composition, texture, grain size etc. The Youngs' Modules is in the range $0,03-0,05 \cdot 10^6$ MPa. The gneisses are mostly quite massive with relatively few fracture zones etc.

In the middle of the mountain range the overburden is about 1100 m.

The tunnel cross section is nearly semi-sircular (Norwegian Road Authority Standard) and the blasted area is 50 m^2. (fig. 5).

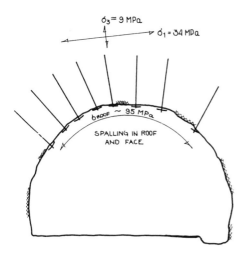

Fig. 5   Cross section of the tunnel

The tunnel was driven from both sides. Conventional drilling and blasting tecniques were used with an old Gardner-Denver 4 boom compressed air jumbo from the Høyanger side, and an Atlas Copco 3 boom hydraulic jumbo from the Lånefjord side.

During planning of the tunnel rock stress problems were anticipated in the tunnel sidewalls under the maximum overburden parts. However, when the tunnelling started from the Høyanger side, quite heavy spalling occured in the tunnel roof only about 200 m from the valley side. Rock stress measurements carried out 800 m from the mountain side showed a dominating horizontal maximum principal stress oriented normal to the tunnel axis (and parallell to the mountain range). This tendency was later on verified by measurements 1450 m from the surface on the Høyanger side and 700 m from the surface at the Lånefjord side.

The highest major principal stress σ_1 was 34 MPa, while the minor principal stress at the same place was 8 MPa.

A simple analysis based on the measured stress data and the Kirsch formulas for the stress distribution around a circular hole was carried out. At the measuring site this analysis showed a tangential stress in the tunnel roof of approx. 95 MPa, while the tunnel wall was under slight tension. Compared to an uniaxial strength of 60-200 MPa, this explains the spalling in the roof.

Assuming that the vertical stress was governed by overburden and that the horizontal stress was constant, similar analysises were carried out at overburdens 900 m and 1100 m. These gave the following tangential stresses:

Overburden 900 m: σ_{roof} = 77 MPa

σ_{wall} = 38 MPa

Overburden 1100 m: σ_{roof} = 73 MPa

σ_{wall} = 55 MPa

This shows a decreasing roof stress and an increasing wall stress as overburden increases. From this simple analysis and the topographical and the geological conditions the following "stress problem forecast" was given:
1) As overburden increases the spalling tendency in the roof will be less pronounced. However, the tangential stresses are still so high that

some problems may occur.

2) Under the maximum overburden the stresses in the walls have increased so much that spalling also may occur here.

During the three years working periode afterwards, the forecast has turned out to be in fairly good agreement with practical experiences. As mentioned above, subsequent rock stress measurements verified the virgin stress pattern indicated by the first measurements.

The rock bursts and spalling created serious disturbances in the tunnelling advance.

At the Lånefjord side the rock was jointed the first 500-600 m from the valley side. From that point on considerable spalling from the roof occured. As the tunnel advanced the intencity increased. Heavy spalling and rock burst occured from the roof and face. The maximum intensity was reached about 2000 m from the valley side, and then decreased until the hole-through about 2800 m from the entrance.

The high intensity of spalling and rock burst at the Lånefjord side was in part caused by frequent changing between massiv granittic gneiss and shistose amphibolite. This may locally cause very high stress concentrations.

At the Høyanger side the spalling and occasionally rock burst in the roof and the face started about 200 m from the valley side and was quite heavy from about 450 m to hole-through. In parts of the tunnel with high overburden some spalling also occured in the side-walls. Some deviations in the pattern occured when crossing fracture zones, crushing zones and lenses of amphibolite.

To control the spalling problems systematic rock bolting was used in the fractured zones (fig. 5). Warm-galvanized 20 mm diameter rebar bolts with a length of 2,4 m were used. The bolts are point anchored with resin cartridges. Triangular roof plates with approx. 300 mm sides are used. When installing the bolts the bolts are not tensioned. The deformation of the rock will eventually put up the necessary tension in the bolt. Pull out tests showed that the threads were torn off at a jacking force of about 18 tonnes. Only a few bolts have been reported to be torn off in practice.

At the Lånefjord side the drilling of the bolt holes was done with a 2 boom pneumatic drill

jumbo equipped with a working platform for the inserting of the bolts. The feeders were specially mounted to drill bolt holes in different direction. At the Høyanger side the drilling was done with hand held jacklegs operated from a hydraulic platform. In periods with heavy spalling and rock bursts no scaling was done during mucking out of the first half of a round. Then scaling and rock bolting was carried out as far as one could reach. Then the rest of the mucking was done. After that scaling and rock bolting was carried out to the face. In the worst zones wire mesh also was used, and sometimes the face itself had to be rockbolted with 1 m bolts.

In periods with more moderate spalling, scaling of the whole round length was done before mucking. Afterwards the bolting was carried out in one operation.

Dependent on the intencity of the spalling and rock bursts, 35 to 70 rock bolts have been used per 4 m round. In the most intence zones wire mesh has been used in the roof (and sometimes also in the face).

At the time of hole-through a total of 55000 bolts had been inserted in the tunnel. Later another 20000 bolts and 26000 m^2 of wire mesh were added.

The cost per bolt was N kr 85 (1 US $ = 7 Norwegian kroner). The cost of the wire mesh was N kr 40 per meter.

Without rock stress problems the driving rate was estimated to 9-11 rounds a week (10 shifts, 40 meters).

Because of the rock stress problems the driving rate sometimes decreased to 5-6 rounds a week (i.e. about the half of the estimated rate).

Fig. 6 shows the cost per meter tunnel as a function of driving rate at the Høyanger side.

With a driving rate of 40 m/week the cost per meter blasted tunnel was N kr 5308 in 1977 (corresponding to N kr 9850 in 1983).

If the driving rate decreases to 12 m/week the cost per meter was N kr 10307 (corresponding to N kr 18400 in 1983).

With a driving rate of 22 m/week which was the average in 1977, the cost per meter blasted was N kr 7000 + bolts N kr 680 = N kr 7680. Compared to the "normal" rate of 40 m/week this

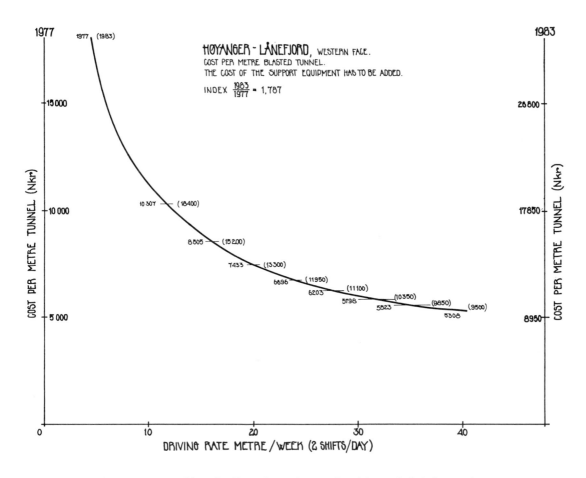

Fig. 6 Tunnel costs as function of driving rate

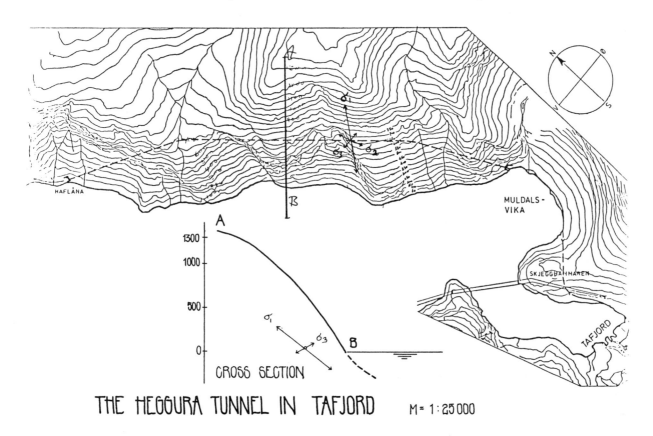

THE HEGGURA TUNNEL IN TAFJORD   M = 1 : 25 000

Fig. 7 Plan and cross section of the Heggura tunnel

means an extra cost of approx. N kr 3800 (1983 prices) due to rock stress problems.

The average rate in Lånefjord was 21.6 m/week. Compared to the "normal" rate the extra cost because of rock stress problems here has been calculated to about N kr 5150 per meter.

The Heggura tunnel in Tafjord

This tunnel represents the "traditional" along fjord type of tunnel. The excavation of the tunnel was started in 1980 and it was operative in 1982. The total length of the tunnel is 5266 m. The tunnel cross section is nearly semi-circular with an area of 39 m^2. Fig. 7 shows the location of the tunnel together with a cross section through the mountain side.

The rock mass consists of different types of gneissic rocks with some amphibolites. The uniaxial compressive strength is in the range 100-250 MPa, and the Young's Modulus is in the range 0.01-0.03 · 10^6 MPa. The gneisses are generally massive with moderate fracturing and jointing. As will be seen from fig. 7 the mountain side is very steep. The maximum vertical overburden is about 700 m, but the mountains behind have heights up to 1800 m. During planning rock stress problems were anticipated due to the "valley side" stress pattern.

Extensive spalling occured about 800 m from the entrance at both sides of the tunnel. As expected the spalling occured in the outside roof corner (towards the fjord) and the inside floor corner (fig. 1).

Triaxial stress measurements in the tunnel verified a typical valley side stress pattern with a maximum principal stress of 25 MPa dipping parallell to the mountain side (fig. 7).

The rock stress problems in this tunnel were solved with a combination of rock bolting and the use of steel fibre reinforced shotcrete. During driving of a 3800 m section of the tunnel 15500 rock bolts (of the same type as mentioned before) were installed, together with 2650 m^3 of steel fibre reinforced shotcrete, i.e. 16 rock bolts and 2.8 m^3 of shotcrete per 4 m round. The shotcrete was sprayed from the middle of the roof to about 2 m above the floor. A remote controlled hydraulic equipment developed by the Norwegian contractor A/S Høyer-Ellefsen was used for the spraying.

The majority of the rock bolts were installed over the same section of the tunnel.

After blasting and scaling, the rock bolts were installed. After blasting of the next round, the first layer of shotcrete was sprayed. After blasting of the third round, the second layer up to 100 mm total thickness was sprayed.

After hole through another 5250 rock bolts were installed, of which 2400 through the shotcrete.

The number of rock bolts per meter tunnels varied very much with spalling intencity. At locations with high, moderate, low or none spalling, 7.7, 4.8, 3.1 and 2.6 bolts per meter were installed respectively. The corresponding driving rate was 31.7 m, 39.5 m, 43.5 m and 52.6 m per week.

The total support cost per meter tunnel were N kr 5633, N kr 4632, N kr 4004 and N kr 3724 respectively (1982).

One year after the finish of the work, the tunnel was mapped in detail. At 150 locations spalling had occured through the shotcrete layer. Usually the shotcrete had very little thickness at those locations. The spalling possibly happened a relatively short time after spraying and the conditions now seem to be stable. A new mapping will be carried out this year to check this.

The use of shotcrete seem to stop further spalling in a tunnel, provided that the layer has a sufficient thickness. Steel fibre reininforced shotcrete has a considerable higher flexability than conventional shotcrete and can take larger deformations without breaking off the surface.

Compared to the rock bolting/wire mesh method used in the Høyanger-Lånefjord tunnel, the shotcrete concept is more expensive, but at the same time considerably faster.

CONCLUSION

The rock burst and spalling problems met in the construction of highway tunnels in Western Norway have been satisfactory solved by either rock bolting/wire mesh or rock bolting/steel fibre reinforced shotcrete. Only those parts of the tunnels surface that are actually affected by

rock burts or spalling are usually supported.
The rest of the surface is without support.
This is in accordance with tradition in Norway,
as most highway tunnels are completely unlined.

Microseismic monitoring in a uranium mine

P. MacDonald B.Sc., M.Sc.
Canada Centre for Mineral and Energy Technology, Ontario, Canada
S.N. Muppalaneni B.Sc., B.Eng., M.Eng., P.Eng.
Rio Algom, Ltd., Elliot Lake Operations, Ontario, Canada

SYNOPSIS

A series of rockbursts, commencing in March 1982, have occurred in a worked out area of New Quirke Mine, Ontario. Mining in an adjacent section was suspended at this time. Prior to recommencing operations, it was appreciated that accurate location and evaluation of seismic events and rockbursts were required.

A 16-channel microseismic monitoring system was installed around the periphery of the affected area, connected to the processing unit on surface. Automatic identification and plotting of events has been achieved by interfacing a microcomputer to the system and developing suitable software. Daily summaries are presented in pictorial form on segments of digitized mine plans. Files of events are also stored for long-term plotting and identification of stress concentrations as reflected by seismic and microseismic activity.

INTRODUCTION

New Quirke Mine of Rio Algom Ltd. is located at Elliot Lake, Ontario as shown in Fig 1. The uranium-bearing conglomerate reefs, part of the Precambrian Matinenda Formation, lay at depths between 250 and 750 m, are 3-6 m thick and dip to southwest at an average 20°. The sequence of reefs and adjacent quartzitic strata are classified as strong, brittle and elastic.

The main reef is currently being mined by stope and pillar method as illustrated in Fig. 2. Levels are developed at vertical intervals of 47 m with stope and rib pillars aligned on dip. Crown and sill pillars are located at the top and bottom, respectively, of each stope. The ore, broken by blasting, is scraped by slushers into

Fig. 1 Location plan of New Quirke Mine.

a boxhole at the bottom of the stope, which is connected to a haulage drift 8 m below the orebody horizon. Extraction ratio varies from 70 to 80%.

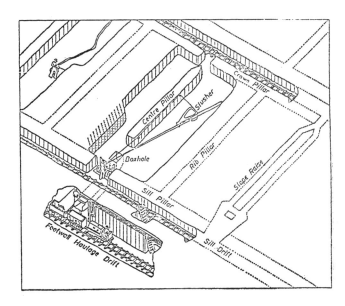

Fig. 2 Stope-and-pillar mining method, New Quirke Mine.

HISTORY OF ROCKBURSTING

In the central area of the Eastern section of the mine, a localized roll in the orebody produced a flat-lying area which proved difficult to mine using slushers. Mechanized stoping was therefore introduced in this block and pillars were realigned at 45° to dip to facilitate scooptram transportation. This trackless area, just below 700 level at a depth of 500 m, was stable during mining operations. However, as mining proceeded in down-dip stopes, effects of increased stress were observed in the pillars. Deterioration associated with a low-angle thrust fault was noted and unconsolidated material was extruded from the plane of the discontinuity.

Stress-related failures in the trackless area pillars were noted in mid 1981, over two years after extraction was completed in the location. Progressive degradation was then observed, including heads broken off rockbolts, fractured pillars, pillar spalling and floor heave, features which were uncommon in New Quirke Mine. Damage was initially restricted to the trackless rib pillars and the sill pillars of some adjacent up-dip stopes in the succeeding nine months, the area of disturbance radiated outward and strata noise was heard by mine personnel in the first quarter of 1982. On 10th March 1982, a small rockburst occurred, probably in the sill pillars above the trackless area, and this was followed by many more in the next months. The violence

with which energy was released by some events was evident in that not only had some rock been propelled across roadways or into the middle of stopes, but also that some rockbursts were felt on surface. The East Canada Seismic Network, operated by Energy, Mines and Resources Canada, Earth Physics Branch, detected certain events, the largest possessing a Magnitude of 3.0.

MULTICHANNEL MICROSEISMIC MONITORING

As a result of the rockbursts in the Eastern portion of the mine, preliminary investigations were conducted employing single channel monitoring instruments. A short period seismograph and a wideband microseismic monitor were used in turn to prove the feasibility of microseismic monitoring at New Quirke and to provide mine management with a continuous record of strata noise. Production was suspended in the vicinity of the disturbed area at this time. Microseismic activity declined gradually from a peak during March 12-14th 1982, however, a more accurate assessment of the stability of the strata in and around the trackless area was required before production could be resumed in the stopes to the south. The following objectives were identified in order to specify the type of system required:

  i)   The source and energy of each seismic event should be determined;

 ii)   The rapid transmission of relevant information to mine management;

iii)   To discover any relationship between strata noise, pillar deterioration and mining activities;

 iv)   The recognition, if possible, of any trends in microseismic activity preceding rockbursts;

  v)   By these and any other means, assess the stability of pillars and hanging wall.

Based on these criteria an ElectroLab 250-MP microseismic system, shown in Fig. 3, was selected for installation at New Quirke Mine. An array of sixteen geophones, located underground as in Fig. 4, are used to detect microseismic events, from which the signals are transmitted to a central analysis unit on surface. There, two microprocessors automatically identify and locate events on the basis of the arrival time of the vibration at a minimum of five detectors. The

Fig. 3 ElectroLab 250-MP microseismic system.

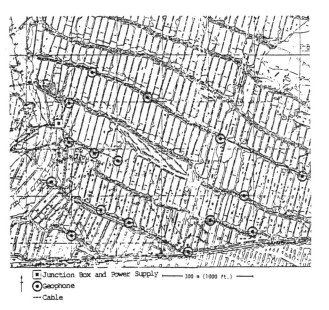

Fig. 4 Distribution of 16-channel geophone array.

computation is performed by a direct method, solving linear equations using geophone coordinates retained in memory. A maximum of four solutions are calculated, resulting in a choice of spatial location and seismic velocity when six or more geophones are activated. Should only five geophones have been triggered, an assumed velocity of 5950 m/sec (19500 ft/sec), is used, determined by calibration tests and stored in memory.

A microcomputer is used as a terminal for input/output and for data processing beyond that incorporated in the system. A menu-driven suite of programs was written, in interpreted BASIC, in

two categories; (a) interface programming to receive, process and store event parameters, and (b) graphics software to enable computer plotting of event locations on mine plans. Fig. 5 summarizes the operation of these programs.

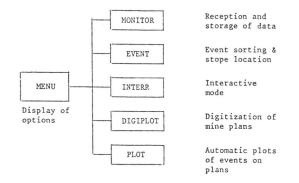

Fig. 5 Operation of program. 'MENU'.

Automatic selection of the closest source from the four possible locations computed by the 250-MP is achieved by making two assumptions. Firstly, that 5950 m/sec is representative of the seismic velocity in all directions and that the events occur within the orebody. The latter assumption is often corroborated by observation of pillar damage after the event. However, lack of access to strata outside the tabular orebody has resulted in an array that is almost planar, the dimensions being approximately 1000 m x 600 m with the maximum dimension perpendicular to the orebody being 25 m. Consequently, source locations are calculated with relatively poor vertical resolution. The best location is therefore considered to be that with the lowest product of the deviations from (i) the assumed velocity and (ii) elevation of the orebody at the calulated plan position. Horizontal location accuracy varies, but is generally ± 15 m within the array, and ± 50 m up to 150 m outside the array.

Discrimination against the inter-shift blasting is achieved by input of blasting times to the program and an editing function facilitates event addition to, or deletion from daily files. Results are then available in tabular format. In addition, the high-resolution graphics capability of the microcomputer enables microseismic events to be plotted on digitized mine plans. These can then be viewed on the VDU or output to a dot matrix printer. Fig. 6 illustrates a typical daily result for one part of the mine.

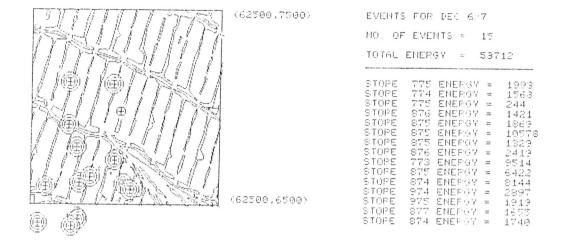

Fig. 6 Typical daily summary of events in one sector of mine, 300 x 300 m.

Files of events are employed for the generation of monthly plots of seismic and microseismic activity during the preceding three months. Cumulative event frequency and total energy are calculated for each 30 m x 30 m segment of the area monitored by the geophone array. This data is then processed by a large computer in Ottawa and overlay plans at 1/2400 scale (1" : 200 ft) are generated on a plotter. A representation of these plans is shown in Fig. 7. This pictorial information is used to analyze trends in the locations of seismic events and rockbursts.

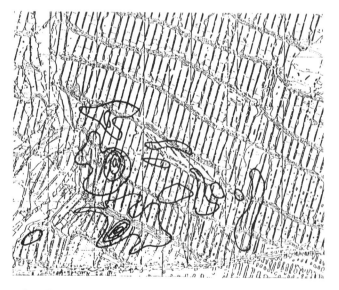

Fig. 7 Event frequency contours for the period August to October 1982. Contour interval equivalent to 0·002 events/m^2.

Concentrations of events or energy are considered to be indicative of zones of stress concentration. Mine management can then identify trends in stress redistribution and relate these to mining operations and the safety of underground personnel.

CONCLUSIONS

The effects of high extraction ratio, pillars aligned at apparent dip, a thrust fault and a localized orebody roll combined to initiate a rockburst zone in an area of New Quirke Mine worked out two years previously. Installation of a microseismic monitoring system has resulted in accurate locations of seismic events within the array. Furthermore, acceptable accuracy, ± 50 m, is achieved for events occurring in the orebody over an area of 1.5 km^2, this being verified by subsequent observation of rockburst sites. Most appear to have occurred in pillars, with some events in haulage drifts below the orebody.

A degree of uncertainty in the location of events in the hanging wall and footwall is introduced due to the planar form of the array. The system is currently being expanded to 32 channels, including two geophones on surface, 500 m above the affected area. It is anticipated that this will improve the vertical resolution of the source determination.

The rapid location of events allows mine management to assess the information before work crews go underground. This data can be displayed on the microcomputer VDU, or output to a printer in pictorial form. In addition, event parameters are stored for further processing, including production of three-monthly moving averages of seismic activity. These results, in conjunction

with convergence measurements and visual
observations confirm that resumption of mining in
the vicinity has had no significant effect on the
major disturbed area.

Some objectives remain to be fulfilled,
particularly relating microseismic activity and
other factors to develop a rockburst-warning
scheme. Furthermore, certain aspects concerning
the short- and long-term stability of the Eastern
section of New Quirke Mine have yet to be determined. Research is continuing in these areas so
that the experience of rockbursts here can be
applied to other mines in the area.

Microseismic detection for Camborne Geothermal Project

A.S. Batchelor C.Eng., M.I.M.M.
R. Baria C.Eng., M.I.E.E.
K. Hearn B.Sc.
Camborne School of Mines, Camborne, Cornwall, England

SYNOPSIS

One of the most promising techniques for monitoring the effects of hydraulic stimulation or fracturing has been derived from traditional methods of locating rockbursts. The use of microseismicity has been shown to be useful for defining the shape and orientation of stimulated regions.

The experiment described in this paper is believed to be the first of its kind with an on-line location system mapping the stimulated zone during injection. The operation was undertaken as part of the Camborne School of Mines Geothermal Energy research programme. Many thousands of events were produced, located and presented as maps to enable the interpretation to be conducted on-line.

The results presented in this paper consist of a series of microseismic maps relating to different stages in the hydraulic fracturing. The fracture front can be seen to grow with time and change orientation. These results represent a considerable step forward in the state of the art of monitoring hydraulic stimulation and have been an essential tool in understanding the geothermal reservoir.

INTRODUCTION

This paper is not concerned with rockbursts in the normally accepted sense of the word. However, by stretching its context to include induced seismicity related to exploitation of a geological resource it can be considered to be related to the subject. Contrary to the detrimental effects of mining related rockbursts, the induced seismicity in a hot dry rock geothermal system is the only means by which the growth of the geothermal reservoir can be mapped. The Camborne School of Mines has been working under contract to the United Kingdom Department of Energy and the European Economic Commission on the problem of exploiting such reservoirs to remove heat from rocks with insufficient permeability to enable the pore fluids to be extracted. This has now become known as the hot dry rock process.

Essentially, the process consists of drilling multiple pairs of wells with a separation of the order of 350 m into heated basement rocks at depths of 4-6 km. With judicious orientation and deviated drilling, the geometry of such systems should be able to exploit any natural jointing that may be present. The method of interlinking the wells is by massive hydraulic stimulation using several thousand horsepower in a similar fashion to techniques for stimulating oil and gas wells.[1-5]

The field programme described in this paper took place in two 2200 m deep boreholes purpose-drilled entirely in granite.[6-10] Microseismic events, generated by the stimulation, were detected on a three-dimensional sensor network and processed on-line to determine their location. During the full operations, 30 000 events were detected and approximately 15% produced interpretable signals and were located accurately.

THE PROBLEM

Any attempt at the direct measurement of the size of the induced structure must rely on one or more of the following techniques:

i   Measuring the disturbed strain field and inferring the shape of the dilated region causing such a disturbance

ii Locating microseismic emissions generated by the stimulation process

iii The application of geophysical techniques to determine the post-stimulation disturbance.

Theoretically, the most satisfactory way of undertaking the measurements is to determine the disturbed displacement field created by the operation. However, the impracticality of making displacement measurements at depth from just two wells means that the observations are restricted to the surface with possible spot measurements in adjacent wells. Tiltmeters have been used in this role with the tilt field being integrated to calculate an absolute displacement. From the calculated displacements, the geometry of the disturbing structure has been inferred. Applications of this technique have been described.[11-13] The difficulty of characterising the rock mass and original geometry lead to considerable ambiguity which reduces the usefulness of the technique as depth increases. A detailed preliminary study on the application of tiltmeters for this specific project showed that the expected tilts were less than 4 microradians with a variation of only 4 mm vertical deflection across 3000 m.[14] In mining applications tiltmeters offer a definite advantage over microseismic events because disturbances may develop more slowly. The authors have not found references to systematic tilt monitoring underground in mines but it has certainly been attempted for subsidence monitoring.

Microseismic emissions generated by the stimulation process are believed to be the product of both the tensile fracturing of intact rock and the shearing of fracture or joint surfaces misaligned to the local principal stress field.

There are several case studies of attempts to map fractures from boreholes using microseismic emissions[15,16] but, probably, the most successful general application of the technique is to locate induced seismicity caused by mining where it is now virtually routine.[17-19]

The third technique of measuring the created disturbance by resistivity, gravity and other geophysical methods has been tried[20] but there is insufficient data to permit an accurate evaluation of the problems in the interpretation.

The limited amount of published experimental fieldwork on this subject showed that the microseismic location system did offer the best chance of success. Considerable reference was made to the data available from the hot dry rock research programme at the Los Alamos National Laboratory, New Mexico.[21,22] It was concluded that microseismic emissions would be produced in regions where significant pressure changes were generated in both the in situ pore fluid and injection fluid. It is not possible to map the region stimulated by the injection fluid alone because the pressures are continuous in the fluids.

QUANTIFYING THE PROBLEM

The two mechanisms for generating microseismic emissions produce very different ground motions. In order to predict the magnitude of the deflection, the emissions from shear sliding on surfaces was treated as a friction plane problem with uniform radiation. The results indicated that particle velocities of approximately 0.1 mm/s at 1000 Hz could be expected at a radius of 300 m. Similar sized, fluid-driven tension cracks gave a similar frequency but a peak particle velocity of less than 0.001 mm/s, a reduction of two orders of magnitude.

The test site is located on a single granite pluton which was expected to be uniform in nature from the ground surface to at least 6000 m. Cross-hole seismic work had been undertaken to depths of 300 m which showed that the attenuation was very low below a depth of 50 m. A series of calculations evaluating the expected signals at radii of 1800-2500 m showed that the particle velocity at the surface would reduce from the 0.1 mm/s to 10^{-3} mm/s with a predominant frequency of 200 Hz. The background noise levels were measured and were found to be one to two orders of magnitudes less than the anticipated signal.

The preliminary conclusions[23] were that the shearing events were likely to be detected but typical tensile or fluid-driven cracks would be extremely difficult to identify because of the much lower signal strength. Tensile events could not possibly be detected on the surface because their incident particle velocities were less than the background noise.

Theoretically, a single three-axis orientated acclerometer package can be used to determine the location of any event by measuring the azimuth of the first arrival and the difference between the primary and shear arrival times. However, the influence of borehole wall motion, clamping, and the directional ambiguity of only one sensor location will produce a misleading result. Another significant problem is ensuring that the different responses between the axes of the triaxial package can be normalised to remove the uncertainty from the azimuthal measurement. It was concluded that three such units should be used to eliminate this tool error.

A hot dry rock doublet consists of two wells and it was originally planned to install the multiple accelerometer packages at intervals of approximately 50 m in the eventual production well. Additional control on the solution derived for the location can be made by considering the rock mass as a homogeneous, isotropic solid and calculating the location from the arrival times at other sensors. A simple analysis of the geometry shows that five sensors are required to produce a single event location based on the first arrival data only. Although the multiple stations in a single plane produce a very stiff set of equations for the location, the ideal set-up was to add two single-point sensors in addition to the multi-axis units in the sensor string at the observation well. These five

arrivals would be used in conjunction with a surface network at radii of 1-2 km from the wellhead to give good epicentral controls.

A detailed evaluation in conjunction with Soil Mechanics Limited and Wimpey Laboratories Limited showed that the engineering difficulties of producing the multi-sensor borehole sonde, approximately 400 m long, were insuperable in the timescale of the operation. Furthermore, there was no field verification of the amplitudes expected from the experiment and it was decided, therefore, to abandon the multi-axis probes in favour of omnidirectional sensors. This removed any chance of conducting hodogram and source mechanism identification but would enable first point arrivals to be used provided the signals were received on the surface and downhole networks. Seismic signals from the explosive stimulation at the bottom of a 300 m borehole were used to estimate the attenuation factor Q for the surface rock mass. This was found to be in excess of 50. The determination of Q for depths greater than 200 m is under investigation and the initial calculations suggest a value in excess of 100.

The final geometry, therefore, was that a string of five hydrophones at approximately 75 m intervals were installed in the observation well which was approximately 300 m away from the injection well. On the surface, vertical axis accelerometers and seismometers were grouted in boreholes 200 m deep at radii up to 2 km away. In addition a network of seismometers and accelerometers conveniently located on the granite outcrop were also incorporated into the network. Signals from the remote outstations were relayed to site by radio telemetry whilst the sensors actually on site and in the well communicated via cable. Two first arrival location algorithms, SPAM and EMI,[24,25] were tested against synthetic data to evaluate both their numerical stability and their inherent accuracy. EMI was the most commonly used algorithm when downhole and sub-surface sensors were operational. EMI is also able to determine the mean compressional wave velocity for a location when there are six or more sensors involved. SPAM was predominantly used when only the subsurface net was available for locating microseismic events.

EQUIPMENT LAYOUT

Unfortunately, during the deployment of the five-point hydrophone string, two sensors were lost in the well and the experiment then ran with the remaining three hydrophones.

Figure 1 shows the sensor positions in relation to the wells. The hydrophone string ran in purpose-designed carriers on a seven-core conventional logging cable whereas the surface units were packaged in tubes and grouted in. The final surface sensor consisted of vertical axis seismic accelerometers suitably stacked and amplified to produce the required sensitivity. The seismic accelerometers were high quality commercial units but the packaging and cabling were purpose designed. Figure 2 shows the data acquisition system in use. It consisted of a recording system to record the raw data on a

Figure 2 The multichannel automatic data acquisition system as installed on site

magnetic tape and a multichannel transient recorder via filters was connected in parallel.

The transient recorder was triggered on receipt of the first arrival and subsequently 8 kbytes of information were digitised for each signal. This very considerable volume of data from 13 channels was transmitted directly to a VAX 11/780 computer for disc storage and preliminary analysis. The signals were processed by purpose-written software and the locations displayed on a high resolution graphics terminal. Figure 3 shows a block diagram of the data path.

Each signal was catalogued by its absolute arrival time and given a unique event number. A pre-processing software package was used to remove most of the uninterpretable signals prior to examination. The problem of discriminating between locatable events, background, environmental or other extraneous noises was overcome by using a combination of amplitude, duration and arrival sequence analyses.

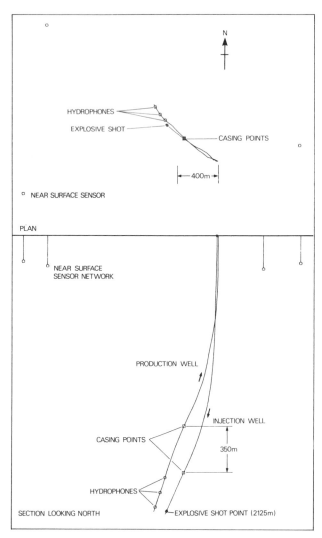

Figure 1 Layout of wells showing sensor positions

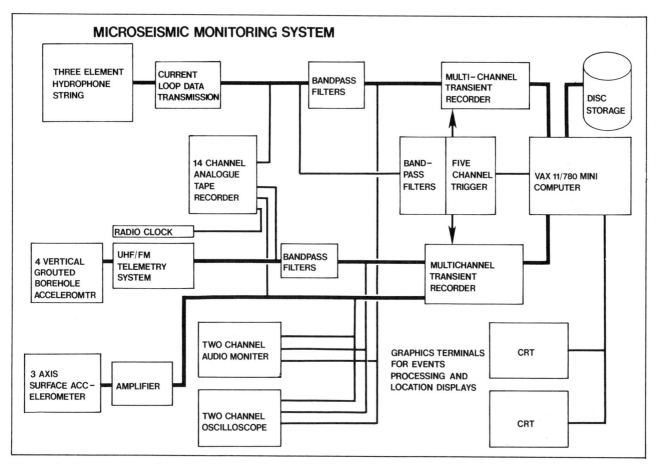

MICROSEISMIC MONITORING SYSTEM

Figure 3 The data acquisition schematic

Unfortunately, the signal-to-noise ratio of the smaller signals was such that manual timing was essential to ensure correct location for the onset of the signal. The larger events certainly had signal-to-noise ratios that were good enough to permit automatic timing.

Once the system was in place it was used to locate a deep explosion and, eventually, a series of calibration shots was run. These measurements show that the assumption of isotropic behaviour within the rock mass was invalid. The average primary wave velocity was 5.73 km/s (53 μs/ft) but it ranged from a value of 5.6-5.9 km/s (51-55 μs/ft) and, moreover, the variation was repeatable from sensor to sensor. The location algorithms were not adjusted but time corrections were applied in the processing software to minimise the location error of the known shot positions. The system was installed and operational by the end of September 1982 for a total budget of approximately $1.5 million. The actual stimulation experiment took place in the period 4-9 November 1982 but seismic monitoring

has continued on-line since that time during production.

THE RESULTS

Figure 4 shows the pressure and flow rate for the period of the main fracturing operation. A peak pressure of 14.2 MPa (2060 lb/in²) was reached approximately 2.5 hours after pumping commenced. The head loss down the wellbore was less than 0.1 MPa (14 lb/in²), making the reservoir breakdown pressure 34.2 MPa (4959 lb/in²) at the bottom of the well compared to a minimum horizontal stress of 30 MPa (4350 lb/in²) and an estimated confining stress on the most favouraby orientated natural joint of 35-38 MPa (5075-5510 lb/in²).

Intense microseismic activity (12-20 events per minute) commenced as the pressure rose in the test but it was not until 12 hours after the start of pumping (3900 m³ (1 million US gal) injected) that the fluid level started to rise in the observation well. Within nine hours it was flowing to surface and the observation well was

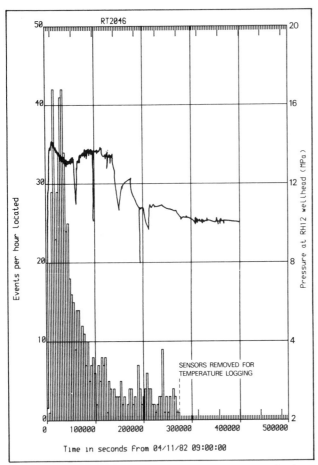

Figure 4 The wellhead presssure during stimulation correlated with the located event rate

shut in. Pumping at an average rate of 82 l/s (1300 US gpm) continued for 38 hours with a total injection of 12 000 m³ (3.2 million US gal).

The flow rate was then reduced because the fluid storage on site could not continue at the same high rate and eventually the experiment was terminated after 120 hours and 18 500 m³ (4.8 million US gal) had been injected. This point has been taken as the formal end of hydraulic stimulation although experiments to circulate and produce the HDR reservoir continued.

Prior to stimulation commencing, the injection well had been treated with a purpose-designed explosive charge[26,27] to reduce the wellbore-to-fracture impedance during the high flow rate stimulation. The effectiveness of this treatment can be seen in Figure 5 which is a series of spinner logs taken during stimulation. Increasing shift across the diagram shows that an ever-increasing percentage of flow is leaving the well at the shot zone at a depth of 2125 m. After 10

hours, it can be seen that half of the flow leaves the well at that point, the remainder being split between the three other well-defined exit points.

The intense signal rate, averaging one every six seconds, produced severe demands on the instrumentation and computing facility. However, it was possible to process the data on a continuous basis. Typical signals are shown in Figures 6a and 6b with their spectra and it can be seen that the predicted frequency of 200-300 Hz is immediately apparent. The acceleration shown is 0.15 mg which corresponds to a particle velocity of 0.0012 mm/s, which is in good agreement with the initial predictions. Figure 7 shows the seismic events received within the first four hours of injection. This view is drawn normal to the projected plane of both wells and the grouping can be seen clearly. Figure 8, which is the view in the plane of the wells, confirmed that the events were located within 50 m of the well. Some events can be seen along the axis of the well and these correspond to the flow exit

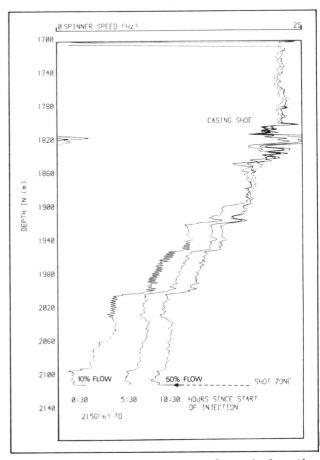

Figure 5 Three repeat spinner logs during the first 10 hours. Note change in flow distribution with time

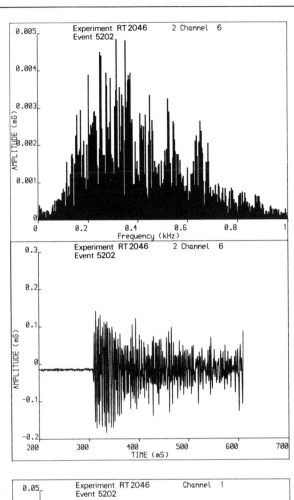

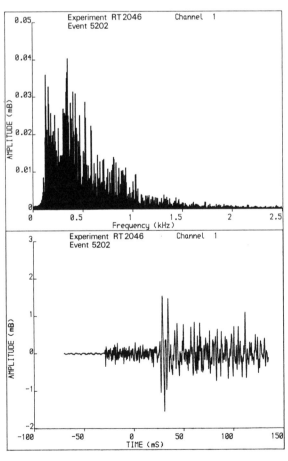

Figures 6a & 6b Typical events with their spectra

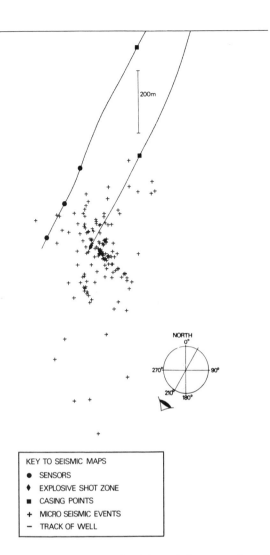

KEY TO SEISMIC MAPS
● SENSORS
♦ EXPLOSIVE SHOT ZONE
■ CASING POINTS
+ MICRO SEISMIC EVENTS
− TRACK OF WELL

Figure 7 Events during the first 3 hours, viewed normal to wells (with key)

points on the spinner log shown in Figure 5. Principally, however, the clustering is adjacent to the shot zone although slightly offset from it. This period is during the build-up to the maximum seismic event rate and before the peak pressure on the well was reached.

Figure 9 shows the next eight hours and the definite progression away from the well can be seen. The tendency for the events to grow downwards was observed in this period but they are still clustered in the plane of the wells. This can be seen in Figure 10. The events seem to be clustering near the base of the well with little evidence of activity near the other flow exits. The data for the next 20 hours shows that the events are still occurring in the same area and these are shown in Figure 11. The corresponding edgewise view was very similar to Figure 10 and has not been presented in this

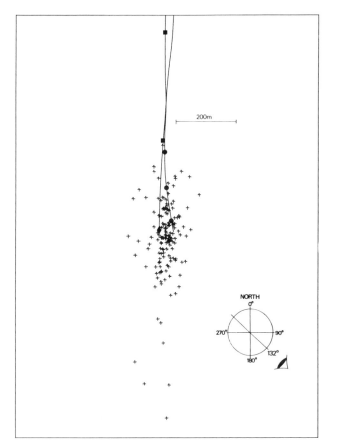

Figure 8 Events during the first 3 hours, viewed in the plane of the wells

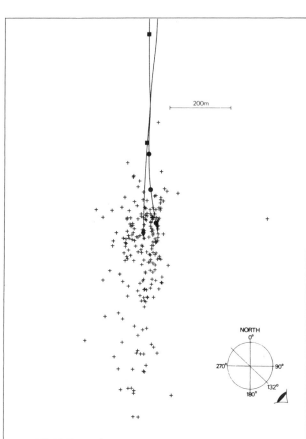

Figure 10 Events from 4th to 11th hour, view - in plane

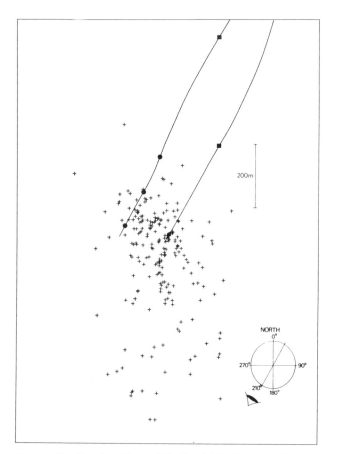

Figure 9 Events from 4th to 11th hour, view - normal

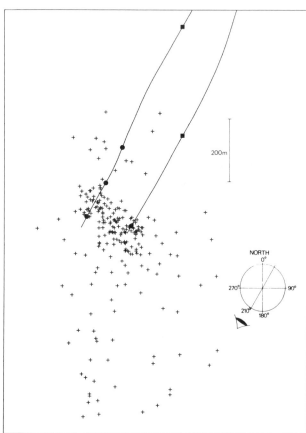

Figure 11 Events from 12th to 27th hour, view - normal

sequence. The overall pattern does show that there was a progressive growth away from the wells. The marked downward development is evident. Note that the region near the shot zone has become aseismic. Most of the seismicity has migrated away from the injection well towards the producing well. Virtually no seismicity was observed throughout the sequence at the other flow exits in the injection well.

The composite view of the entire sequence is presented in Figure 12 with its corresponding in-plane view in Figure 13. The confinement by the natural stress field is obvious. In the plane of the wells, the maximum stress direction, the stimulated zone has grown to approximately 400 m in length but it is over 700 m high and more than half of the region lies below the stimulation point. The in-plane view shows that the majority of the events are contained in a zone only 50 m wide. The downhole sensor string was removed from the observation well after 72 hours to permit flowmeter spinner logs to be run in the production zone. The well was found to be producing at a depth of approximately 1700 m, which was a zone not associated with the substantial number of microseismic events. The log also showed water entering the base of the well at slightly above the well equilibrium temperatures, showing that water was, indeed, following the microseismic event concentration at the base of the production well.

During the subsequent 20 days of circulation, intense microseismic activity continued and the stimulated zone developed further. Figure 14 shows the in-plane view of the events during this period. Note the lack of events near the well.

The microseismic measurements have been pre- and post-calibrated and it seems likely that the errors in the plots are less than $\pm$ 20 m. This error may account for the apparent width of the in-plane view but it does not account for the size in the principal direction. The spread in the events occurring below the well does have a component of numerical instability, but this is believed to be less than $\pm$ 50 m during the sequence presented.

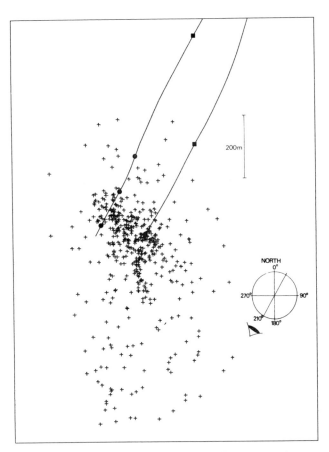

Figure 12 All events from first 27 hours, view - normal

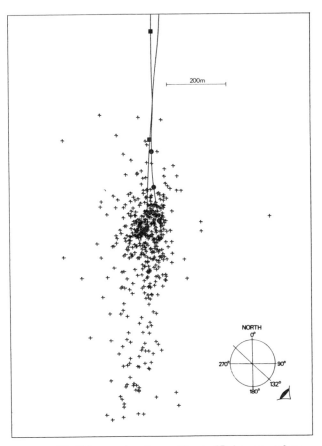

Figure 13 All events from first 27 hours, view - in plane

155

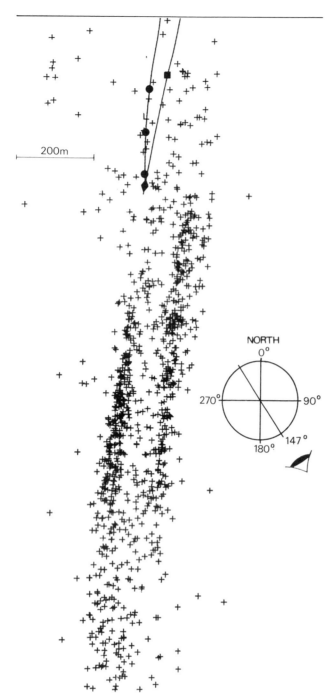

Figure 14 All events from next 27 days, view - in plane. Note scale change

THE INTERPRETATION

The events detected were both similar in amplitude and frequency to those anticipated for shear events in the original design of the system. The frequency of 200-400 Hz compressional wave implies a source length of approximately 10 m and the significant amplitude of the signals at sensor radii of up to 2 km meant that the events were related to reasonably energetic occurrences. The downward growth of the reservoir and the lack

of seismicity at points known to be accepting a substantial flow rate during injection and known to be producing zones in the production clearly implied that the seismic mapping was not defining the entire stimulated region. However, the logging in the production well in the region where seismicity was detected did show that flows were occurring at those points and it has been concluded · that the presence of seismic events does indicate that there is a hydraulic connection to that point in the injection well, but not all hydraulic connections will be seismically active.

The definite gap between the two 'sides' of the structure shown in Figure 14 has been interpreted as the main flow with the seismicity caused by the permeation at right angles to the zone. The growth of the region can be matched by a bulk permeability of only 10 μd (10^{-17} m^2) which is similar to the unstimulated granite in situ. The downward growth of the reservoir was of particular interest because of its unexpected nature. It is believed that this phenomenon can be explained with reference to the measured state of stress and the jointing. Figure 15 shows the principal joint directions and local stress field superimposed on each other. The downward growth is probably a result of the high horizontal shear stresses acting on the vertical joints, striking

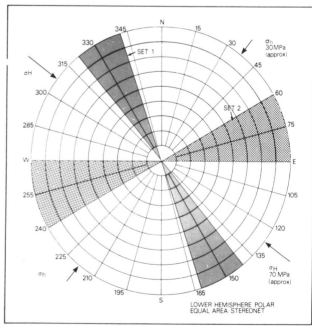

Figure 15 Stress field and joint directions at 2000 m

at about 10-30° from the maximum stress direction. A simple model for the critical hydraulic pressure required to initiate shearing was developed by Pine[28] and he showed that, if a critical pressure at an intersected joint was below that of the equivalent critical pressure at a different elevation, then the growth would be downward. This is a completely opposite effect to that predicted by the conventional theory that if the minimum stress gradient is greater than the hydrostatic gradient, then the fracture growth will be upward. In this case, it is the shear stress gradient that is sub-hydrostatic and the downward growth was caused by shear. Taking Pine's analysis further, it can be shown that downward shearing should not occur above depths of 1400 m. The illustrations show that virtually no seismic activity occurred at those depths, yet substantial flows were detectable. Conventional hydraulic stimulation assumes that fracture or joint opening will be normal to the minimum principal stress. This is believed to be the first time that measurements have confirmed that stimulated structures in strong rocks will grow by an alternative mechanism if natural joint direction is offset to the stress directions.

The corollary of this conclusion was that shear growth will occur at pressures much below that of the minimum stress and that the principal mechanism of the far-field growth will be by shearing activity, unless the pressure gradient is contained by the use of a viscous gel or pressure reductions at the production well.

The events were reprocessed and their spectra analysed to determine whether there was a class of events with frequencies of approximately 1000 Hz. Several high frequency events that did not occur on the surface network were found on the downhole sensors. These occurred early in the stimulation sequence but were soon overwhelmed with the larger amplitude shear events.

In May 1983, a 195 1/s (73 B/min) injection using a conventional fracturing fluid, Dowell Schlumberger YF8 fluid, and 5500 hydraulic horsepower, was undertaken. No seismicity whatsoever was detected, yet a television log run after the viscous fracturing operation showed the

well actually to be fractured over two hundred metres (Figure 16). Damage material recovered from the well showed that substantial fracturing had taken place during the stimulation. The plane of the fracture growing from the well was exactly aligned with the maximum stress field and therefore there was no shear motion present across the structure. This experiment has been taken to confirm the view that shear mechanisms produce detectable seismicity, yet the principal fracturing operation and flow zones more closely aligned to the stress field are much less seismically active. The only method that can be used to locate the fluid-driven tensile cracks that are generated during stimulation is the use of high sensitivity, oriented, multi-axis accelerometers clamped in the borehole. The experimental data obtained showed that these high frequency, low amplitude events could not be detected on the surface, 2000 m away, even in the ideal circumstances of this experiment. The lack of seismicity during the conventional hydraulic fracturing process also implied that the technique is not universally applicable.

Figure 16 Photograph of a self-propped fracture 5 mm wide at 1578 m deep. Full field of view is 100 mm

CONCLUSIONS

The following conclusions may be drawn:

● The majority of the detectable seismicity generated during the hydraulic stimulation was caused by shear events with a characteristic length of approximately 10 m.

- High frequency, low amplitude events were detected but not located on a downhole network only 350 m from the injection point, but the vast majority of bigger events overwhelmed these signals.

- It has proved possible to map the growth of the stimulated region on-line.

- The stimulated zone has been contained in the plane of the maximum principal stress despite one of the growth mechanisms being by shearing action.

- Downward growth of the stimulated zone has been detected and confirmed by production performance and this can only have occurred if shearing was the principal reservoir growth activity.

- This technique can only be one of several diagnostic techniques because the system is "blind" to stimulation occurring normal to the minimum principal stress.

The overall location system can be adapted to any geometry and any number of sensors and is applicable to many geotechnical engineering fields where microseismic activity has engineering significance.

ACKNOWLEDGEMENTS

This work was carried out under contract to the European Economic Community and the United Kingdom Department of Energy, contract numbers EG-D-2-003-UK(N) and E5A/CON/115/173/017. The support and encouragement of the staff of these organisations is gratefully acknowledged.

The 58 full-time staff working on the Camborne School of Mines hot dry rock geothermal programme have all been involved in this experiment to some degree and the support and enthusiastic involvement of all the staff is acknowledged.

The Institute of Geological Sciences Engineering Geology Unit, and Dr D McCann in particular, have been involved with the programme from the start and the authors extend their grateful thanks for the support.

All the views and conclusions expressed in this paper are the authors' personal opinions and are not necessarily supported by either the United Kingdom Department of Energy or the European Economic Community Directorate General for Science and Research.

REFERENCES

1 **Armstead H C:** "Geothermal energy", London: F & N Spon, 1979, 356 pp.

2 **Robinson E S, Potter R M, McInteer B B, Rowley J C, Armstrong D E, Mills R L:** "A preliminary study of the nuclear subterrene Appendix F, Geothermal Energy", Los Alamos Scientific Laboratory, 1971. (LA-4547) 62 pp

3 **Smith M C:** "The potential for the production of power from geothermal resources", Los Alamos Scientific Laboratory, 1975. (LA-UR-73-926)

4 **Garnish J D:** "Geothermal energy: the case for research in the United Kingdom", London: HMSO, 1976. (Energy Paper No 9) pp 66

5 **Batchelor A S, Pearson C M:** "Preliminary studies of dry rock geothermal exploitation in South West England", Transactions of the Institution of Mining and Metallurgy (Sect B: Appl Earth Sci), 88, 1979. pp B51-6

6 **Beswick A J:** "Drilling deep geothermal wells in Cornish granite: Pt 1, Planning and drilling the wells", Geodrilling, No 21, Apr 1983. pp 18-25

7 **Beswick A J:** "Drilling deep geothermal wells in Cornish granite: Pt 2, Directional drilling, motors and casing", Geodrilling, No 22, Jun 1983.

8 **Beswick A J:** "Drilling deep geothermal wells in Cornish granite: Pt 3, Post-drilling operations and future drilling plans", Geodrilling, No 23, Aug 1983.

9 **Beswick A J:** "Well drilling and casing Volume 1 Summary Report", Camborne School of Mines Geothermal Energy Project, 1982. (Report 2-15) 31 pp

10 **Beswick A J, Forrest J:** "New low-speed high torque motor experience in Europe", Society of Petroleum Engineers of AIME, 1982. (SPE-11168) 16 pp

11 **Evans K:** "On the development of shallow hydraulic fractures as viewed through the surface deformation field: Part 1", Journal of Petroleum Technology, Feb 1983. pp 406-420

12 **Evans K F, Holzhausen G R, Wood M D:** "The geometry of a large-scale nitrogen gas hydraulic fracture formed in Devonian shales: an example of fracture mapping using tiltmeters", Society of Petroleum Engineers Journal, Oct 1982. pp 755-63

13 **Wyatt F, Berger J:** "Investigation of tilt measurements using shallow borehole tilt-meters", Journal of Geophysical Research, Vol 85, No B8, 1980. pp 4351-4362

14 **Pine R J, Fernie R A:** "Microseismic fracture location", Report by Golder Associates (UK) to Camborne School of Mines Geothermal Energy Project, Feb 1981. (Report 8051041/2) 39 pp

15 **Cornet F H:** "Study of acoustic and micro-seismic emissions associated with hydraulic fractures", Seminar on Geothermal Energy, Brussels, Dec 1977. Papers. Commission of the European Communities, 1977. (EUR 5920) pp 685-691

16 **Power D V:** "Acoustic emissions following hydraulic fracturing in a gas well", Conference on acoustic emission/microseismic activity in geolgic structures and materials, 1st, Penn. State Univ., June 1975. Proceedings, edited by R H Hardy and F W Leighton. Aedermannsdorf, Switzerland: Trans Tech Publications, 1977. pp 291-308

17 **Salamon M D G, Weibols G A:** "Digital location of seismic events by underground network seismometers using the arrival times of compressional waves", Rock Mechanics, Vol 6, 1974. pp 141-166

18 **Trombik M, Zuberek W:** "Location of micro-seismic sources at the Szombierki coal mine", Conference on acoutstic emission/microseismic activity in geologic structures and materials, 2nd, Penn. State Univ., Nov 1978. Proceedings, edited by R H Hardy and F W Leighton. Aedermannsdorf, Switzerland: Trans Tech Publications, 1980. pp 179-191

19 **Chamber of Mines of South Africa:** "An industry guide to the amelioration of the hazards of rock bursts and rock falls", Chamber of Mines of South Africa Research Organisation, 1977.

20 **International Energy Agency:** "Man made geothermal energy systems (MAGES) appendix original reports", Paris: Organisation for Economic Co-operation and Development, 1979 (unpublished).

21 **Pearson C F:** "The relationship between microseismicity and high pore pressures during hydraulic stimulation experiments in low permeability granitic rocks", Journal of Geophysical Research, Vol 86, Sep 1981. pp 7855-64

22 **Albright J N, Pearson C F:** "Acoustic emission as a tool for hydraulic fracture location: experience at the Fenton Hill Hot Dry Rock site", Society of Petroleum Engineers Journal, Aug 1982. pp 523-30

23 **Soil Mechanics/Wimpey Joint Venture:** "Identification of the hydraulically stimulated region of a hot dry rock reservoir by acoustic emission", Bracknell, Berkshire: Soil Mechanics, 1981. (Report JV2) 38 pp

24 **Dechman G H, Sun M C:** "Iterative approxi-mation techniques for microseismic source location", US Bureau of Mines, 1977. (Report of Investigations 8254) 23 pp

25 **Thorn-EMI Electronics Ltd:** "A general algorithm for sensor point observers", Camberley, Surrey: 1981. (Doc Ref MS/G/113, Iss 1) 18 pp

26 **Batchelor A S, Pearson C M, Halladay N P:** "The enhancement of the permeability of granite by explosive and hydraulic fracturing", International seminar on the results of the EC Geothermal Energy Research, 2nd, Strasbourg, Mar 1980. Proceedings. Dordrecht: D Reidel, 1980. (EUR 6862) pp 1009-1031

27 **Batchelor A S:** "The creation of hot dry rock systems by combined explosive and hydraulic fracturing", International conference on geothermal energy, Florence, May 1982. Papers. Cranfield, Beds: BHRA Fluid Engineering, 1982. pp 321-342

28 **Pine R J:** "Downward growth of fractures", Bi-monthly Information Circular Camborne School of Mines Geothermal Energy Project (13) Oct-Nov 1982. pp 28-35

Analysis of stoping sequence and support requirements in a high stress environment – ZCCM – Mufulira Division

F.M. Russell M.Sc., D.I.C., A.C.S.M., M.I.M.M., C.Eng.
Mine Superintendent, ZCCM Mufulira Division, Zambia
D.R.M. Armstrong B.Sc., M.I.Geol.
Chief Geologist, ZCCM Mufulira Division, Zambia
R. Talbot B.SC., A.R.S.M., M.I.M.M., C.Eng.
Associate, Golder Associates, Seattle, U.S.A.

SYNOPSIS

For ten years, part of the western section of Mufulira was mined without major problems. The area, which was overlain by a river, was mined using post-sandfill stoping methods to prevent hangingwall dislocation. In 1973 and 1974 rockbursts occurred resulting in loss of life and production.

A geotechnical analysis was carried out using a mining simulation program MSIM3D which indicated that a change in stoping sequence was required, together with rock support. Stoping resumed in the area, but in 1977 more rockbursts occurred and a more detailed investigation was initiated.

The analyses were again carried out using MSIM3D this time with modified input data. Results indicated that mining in this area would always be subject to high stress and that subsidence beneath the stream and ingress of water into the workings was not impossible.

It was then decided to divert the stream and resort to a total extraction method of mining, without post-sandfill and using extensive support of development. This has been carried out over the past five years with success, some 6 million tonnes having been extracted with a minimum of rockburst activity.

The geology of the area, the history and mining methods in use, the MSIM3D analyses and the rock support methods are all described.

INTRODUCTION

The first recorded discovery of copper at Mufulira occurred in 1923. Production commenced in 1933 and since then some 200 million tonnes of ore at 3.5 per cent copper have been depleted leaving reserves of about 95 million tonnes at 3.1 per cent.

Production is currently 5.5 million tonnes of ore/year at a mill head grade of 2.1 per cent copper.

Mufulira is located in the Zambia-Zaire metallogenetic province known throughout the world as "The Copperbelt". Originally owned by Mufulira Copper Mines Ltd. the property formed part of the Roan Selection Trust Group, later to become Roan Consolidated Mines Ltd (RCM). Mufulira is now one of the seven operating divisions of Zambia Consolidated Copper Mines Limited (ZCCM), the company which resulted from the merger of RCM and Nchanga Consolidated Copper Mines Limited (NCCM) in 1981.

GEOLOGY

The Mufulira deposit occurs on the north-eastern side of the Kafue anticline and, in contrast to the shale-type orebodies found to the south-west, the host rock for the copper - iron sulphides is metamorphosed arenaceous sediments. The deposit lies on the south limb of a large syncline known as the Mufulira Basin. There are three superimposed orebodies known as the A (upper), B (middle) and C (lower). The C orebody has a strike length of 6 km and is overlain at the east by the A and B orebodies which have strike lengths of 2 km and 3 km respectively. The dip of the sediments averages 45^{0} but can vary from horizontal to vertical. Under the river area the dip is approximately 30^{0}.

Two typical Geological sections are shown in Figure 1.

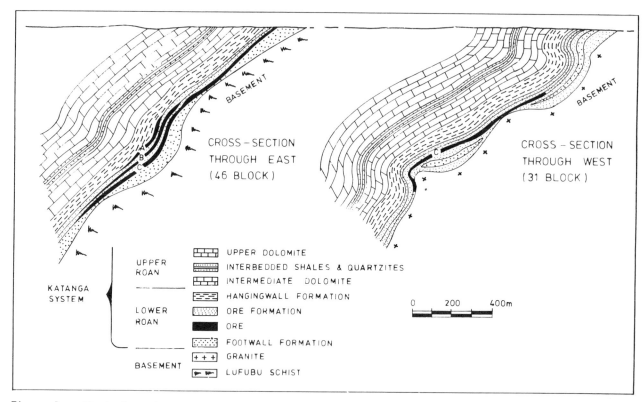

Figure 1.   Typical geological sections through the east and west mining areas

Structure

Faulting and folding are virtually absent at Mufulira and are generally related to Basement palaeo hills of the pre-Katanga topography.

There are three orthogonal joint sets recorded at Mufulira which are predominantly steeply inclined (in excess of 60°) and have not been observed to exceed more than 9 metres in length. The density of major joints (defined as those traceable for more than 3 metres in two dimensions) appears to decrease with depth.

A study carried out in 1974 indicated the results shown in Table 1.

The relative attitudes of the major joint sets and the well bedded footwall quartzites gives rise to occasional wedge failures, especially where the joints are open and in-filled. The more massive rock types, for example the orebody formations and the Inter BC quartzites, exhibit much less bedding and jointing and are more competent.

Rock Strengths

A limited amount of laboratory testwork on core samples has been completed for rock types at Mufulira. Generally the orebody rocks can be classified as very strong (unconfined compressive strength 100 - 200 MPa) with some samples ranging from strong to extremely strong (50 - 200 MPa).

Table 1 -   Summary of Structural Data

Structure	Attitude*	Persistence	Spacing
Bedding	30-70/040-060	Depends on lithology	
Conjugate Joints A	60-80/140-155	3-5 m	2-10 m
Conjugate Joints B	60-75/295-320	4-9 m	1-3 m
Joints	40-50/215-235	4-8 m	~ 25 m

*40/072 denotes a feature dipping at 40 deg. towards bearing 72 deg.

162

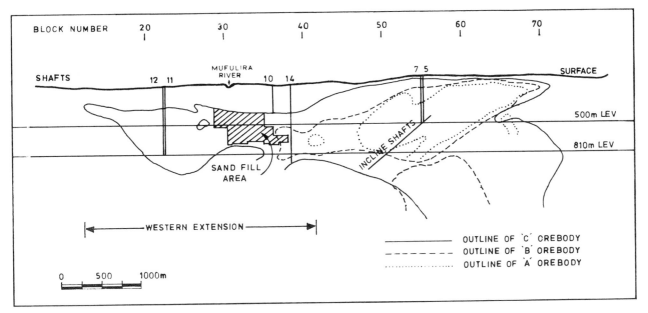

Figure 2.   Vertical longitudinal projection of Mufulira

General

The western part of the mine (including the river area) exhibit certain salient geologic differences from the eastern section which, to a degree, influenced some decisions and contributed to the problems encountered.

1 There is a greater occurrence of cross-bedded sands in the footwall sediments and these, together with structural factors associated with Basement highs (palaeo hills) cause wedge failures and generally a weaker rock.
2 The two upper orebodies (A and B) are uneconomic and attenuated, thereby forming a more competent stoping hangingwall.
3 Certain of the hangingwall dolomitic horizons, which, in the east are aquifers, become impervious in this area. This is due to a decrease in the depth of weathering and a reduction in the solubility of these horizons.

The overall effect is a more competent rock mass in the hangingwall with less planes of weakness which render it less amenable to caving.

MUFULIRA'S WESTERN EXTENSION

In 1947 an exploration programme was drawn up to investigate the west of the mine and in 1952 a borehole intersected mineralisation at a position 750 metres west of what was believed to be the western fringe of the orebodies. The exploration

programme continued and by 1956 sufficient information had been obtained to show that the C orebody extended a further 3 km (making a total of 6 km) along strike. Plans were made to exploit the western extension and by 1963 shafts had been sunk and equipped, underground crushers installed, haulages developed and the area was in full production. (Figure 2)

Mining operations

One of the major problems to be solved in mining the western extension was how to deal with the presence of a river, known locally as the Mufulira Stream, which ran across the orebody. The river has a yearly peak flow of 12 m^3/s, with peak flows of 58 m^3/s possible during floods.

Extensive surface subsidence had occurred at the eastern section of the mine where open stoping and block caving had been carried out. Therefore, even though the orebody in the western section was thinner and deeper than the east, it was thought likely that stoping operations could induce the river into the hangingwall through caving cracks.

Two basic options were examined, one being the piping of the river across the orebody, the other being to sandfill the stopes. Piping of the river was considered expensive in terms of capital outlay whereas sandfilling only gave an increase in operating costs which was then

163

partly offset by the ability to extract the ore with little dilution. The latter option was chosen and stoping sequences were designed so that stoping commenced from one main level and worked upwards to the next main level 76 metres vertically above.

The mining method chosen was an adaption of the sub-level open stoping method in use at the time. Stopes were usually 28m wide along strike with a back length of 35m. Rib pillars were 4m thick and crown pillars 10m thick. Extraction was through grizzleys with scrapers being used in flat dipping areas to scrape down dip in the stope. On completion of the stope, openings were bulkheaded off and classified tailings placed hydraulically in the stope, drainage taking place through filters in the bulkhead.

The delineation of the stoping areas to be sandfilled was by drawing a line at 70^0 to the horizontal from the river position to the intersection with the orebody. This meant that the sandfill area extended along strike with increase in depth. As the river flowed across the orebody at an angle to the strike the sand-fill area was also staggered along strike with increase in depth. This created sandfill stopes under caved stopes and vice versa. The angle of 70^0 was assumed to represent the caving angle, the caving angle being that angle at which indications of both subsidence and tension cracks were found on surface following extraction of the ore below, and was based on observations taken at the eastern section of the mine.

Stoping proceeded down dip, mining upwards from one main level to the next, without major problems. Below the 500m level trackless mining was introduced, the main consequences being higher rates of extraction and larger openings (usually 3.6m high and 4m wide), factors which were to have an influence on subsequent events.

In 1970 Mufulira had the worst mining accident ever to occur in Zambia. Tailings from a dump situated on the hangingwall of the eastern section entered the mine through the cave column, flooding the eastern section with the loss of 89 lives and bringing production to a halt. Following the accident a review was made of the river area. Again the possibility of piping the river was considered but was discounted in terms of cost

Mining in the area above the 660m main level continued upwards to the 580m main level at production rates in excess of 100 000 tonnes/month. The stoping sequence used was one which achieved the maximum tonnage to be mined, the only constraint being that there could not be two adjacent stopes empty at any time, in other words a stope being mined had to have the adjacent stope filled or be intact. No consideration was given to geotechnical constraints although no rock support was used or considered necessary in the area.

In December 1973 the first of a series of rockbursts occurred, resulting in the loss of one life.

In June 1974 the second major rockburst occurred resulting in the loss of four lives.

Following these rock bursts a thorough investigation was carried out by the Rock Mechanics Section of the mine and consultants.

GEOTECHNICAL ANALYSIS
The results of this investigation indicated the following.
1. High rock stresses had been caused by a number of complex, interactive mechanisms; such as; the depth of the top of the orebody, the use of sandfilling with remnant pillars, the unmined pyritic zone to the west and a strong hangingwall rock.
2. Underground observations showed that some failure of old stope pillars had occurred, resulting in local caving of the immediate hangingwall and transfer of stress to the lower mining abutment. The occurrence of this "arching" stress was also evidenced by the negligible amount of surface subsidence which had occurred.
3. On the fringes of the sandfill area abutment stresses had also been induced by the extensive use of caving mining methods to the east and west.
4. Local stress concentrations had been caused by the up-dip sequence of stoping between main levels creating remnant blocks.
5. The foregoing resulted in increasingly high stress being transferred to the strong brittle footwall quartzites which was sufficient to

cause rockbursts in the mining, footwall drives and crosscuts.

The problem then to be solved was how to mine the area safely. To do this some indication of the rock stresses likely to be encountered had to be determined.

Prediction of rock stresses

A series of stress analyses was carried out, first to correlate predicted stresses for the current mine configuration with observed damage underground and second, to predict likely mining conditions for proposed mining methods and stope extraction sequence. This empirical approach was adopted initially because of the limited amount of data that existed for either rock properties, in-situ stresses or measured underground deformation.

Two stress analysis methods were used.
1. A two-dimentional finite element model (FES2D) to determine the stress distribution around stope development drives and crosscuts in the orebody and the footwall.
2. A three-dimensional face element model (MSIM3D); for complex mining layouts in a tabular, single plane orebody, the program determines stresses in the orebody and hanging-wall to footwall convergence and ride over the problem area.[2]

For MSIM3D analysis, a working plan is drawn up by translating the mining geometry (Figure 3)

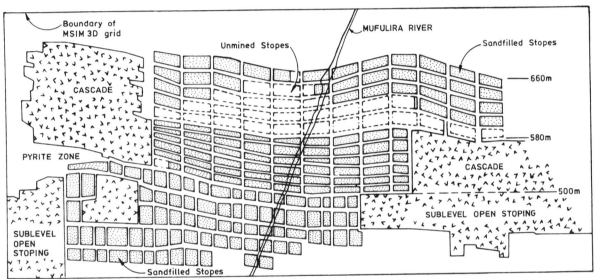

Figure 3.   Horizontal projection showing mining geometry   December 1973

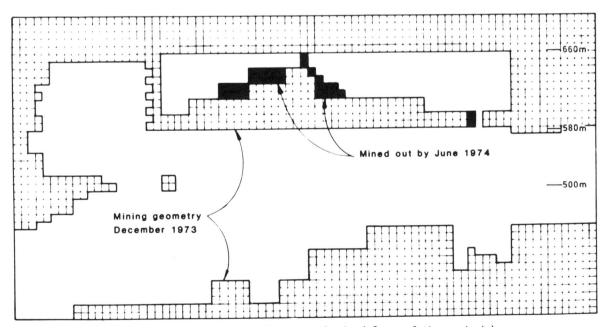

Figure 4.   Grid for MSIM3D, 1973/74 rockburst analysis (plane of the orebody)

to an idealized projection in the plane of the orebody (viewed from above). This is then drawn in the form of a computer grid; analyses for different mining sequences are done by "mining out" grid blocks on the computer (Figure 4). An elastic rock mass is assumed in the hanging-wall and footwall. On the mining plane, material properties can model brittle behaviours,in pillars (characteristic of rockbursts), compressible fill and open, uncaved areas.

The majority of the mining evaluation was done using MSIM3D. The approach taken in the initial analyses was as follows.

1. For a specified mine geometry and material properties the distribution of stress and displacement at the orebody hangingwall was calculated using MSIM3D.

2. For a typical cross section geometry the stress concentration factors around mined openings were calculated using FES2D, for a unit stress applied to the orebody hangingwall.

3. By combining the results from the above two analyses, predicted stresses around the mined openings were correlated with observed rock failure underground; this provided an initial estimate of a "critical" stress value (calculated at the orebody hangingwall) which would be used to identify potential rockburst conditions in analyses of future mining layouts.

In subsequent analyses step 2 was ommitted. It is emphasized that the above analyses were to provide an empirical guide to future mining conditions, rather than absolute stress values. This approach was considered reasonable as the orebody geometry and mining method were not expected to change significantly.

Model input data

Input data required for the analyses included elastic parameters, in situ (pre-mining) stresses, and stress-strain relationships for rock and backfill. Values for Young's modulus (E) and Poisson's ratio (ν) for the orebody and foot-wall rocks had been determined from a limited number of unconfined compression tests on intact NX core samples, and were considered sufficiently precise for the proposed evaluation. In these early analyses a simple linearly elastic stress-strain relationship was assumed for the rock using $E = 3.45 \times 10^7$ kPa and $\nu = 0.2$. For later analyses, a strain softening relationship was

used for isolated pillars of ore where significant crushing failure in the orebody had been observed underground, and for previously sandfilled stoping areas. This relationship was of the general form: a linearly elastic deforma-tion to a peak (failure) stress, linear post-failure deformation to a residual stress, and continued displacement at constant (residual) stress. This is shown in Figure 5. This

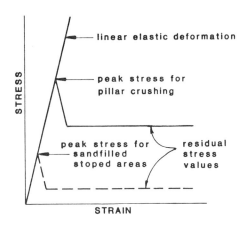

Figure 5. General form of stress-strain relationship for MSIM3D analysis

general form represents the crushing failure, accompanied by load-shedding that was occurring in highly stressed areas of the mine, and also the confining effects of the sandfill and the load bearing capacity of the sandfilled stopes. Little data was available to support specific values for peak and residual stress, and so a sensitivity analysis approach was used. Based on data available from the Copperbelt mines, it was initially assumed that the maximum principal stress was vertical and equal to the overburden load, and the horizontal to vertical stress ratio was 0.5. Although no measurements of in-situ stress values were available, an effort to determine a relative value of horizontal and vertical stress was made in 1976. This was done by taking a series of 300 mm diameter cores (to a depth of 300 mm) instrumented with a circular, 10-pin strain rosette, from the wall of a bored raise. Principal stresses were then determined (using elastic theory) from the measured strains which resulted from over-coring.

A large number of redundant measurements were analyzed and plotted in histogram (magnitude) and stereoplot (direction) form, and a finite element analysis was also used to assess the influence of previous mining activity on the pre-mining stress values. This work indicated that the major stress was near-vertical (72° dip), the intermediate stress was approximately normal to the strike of the orebody, and confirmed that the ratio of the horizontal (intermediate) to vertical stress ratio was 0.57.

Analysis of events, 1973-1975

Analyses were completed using two MSIM3D models. The first to evaluate the cause of the two rockbursts that occurred on the 580/594m levels and assess the effects of additional mining above the 660m level. The second to evaluate the stresses (and potential for rockbursting) for two alternative stoping sequences below the 660m level.

The initial MSIM3D model was based on a grid 1220m on strike by 610m on dip (in the plane of the orebody), bounded on the west by the barren pyrite zone. This was extended in the second model to 1680m on strike by 1200m on dip, to allow mining to the 730m level and to minimise the influence of major caving, east of the sandfill area, on the predicted stresses (MSIM3D cannot explicitly model hangingwall caving). For the two models, minimum grid block dimensions of 15.24m and 24m/12m, respectively were used; in the second model the sandfill stoping area was scaled by a factor of two, to provide more detail and accuracy without ignoring the effect of adjacent caved or supported areas. In these analyses, stress and displacement parallel to the orebody plane were assumed to be small because of the flat dip angle (average 30°) and were not calculated.

Comparison of the initial results with observed damage suggested a "critical" normal stress value 55.8 MPa(8000 psi); the likelihood of further serious failure would be high where predicted stresses exceeded this value. For stope development in the remnant stoping block, maximum shear stresses calculated using FES2D indicated that the inter-drive pillars were highly stressed and possibly failed over approximately half their width. The main conclusions drawn from the initial analysis were as follows.

1. Additional mining in the area above 660m

level could be undertaken provided that all development was adequately supported with fully grouted bolts, though the possibility of additional rockbursts occurring was high.

2. The schedule of stoping should be reversed to a down-dip sequence for all ore remaining below the 660m level.

In 1975, the second series of analyses with MSIM3D were completed to evaluate stoping sequences down to the 730m level and to assess the effect on future mining of leaving an ore remnant on the 580/594m levels (potentially unrecoverable). Figure 6 shows results for the expected mining profile at the end of 1977.

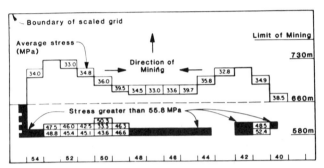

Figure 6.   Predicted average stresses (MSIM3D) December 1977

The results of these analyses indicated that while high stresses would be experienced in stope development down to the 730m level, with the down-dip mining sequence, values would be well below the "critical" stress 55.8 MPa. It was also concluded that further mining above 660m level would not significantly increase stresses on the lower stoping face. It was recognized, however, that mining conditions on the 580/594m levels would be increasingly more difficult as additional stopes were extracted and it was determined that the decision to mine each stope would be based on careful observation of local conditions and anticipated stresses and that monitoring of surface subsidence, local hangingwall caving, and evidence of increased stresses over the whole sandfill area should be continued.

Two major assumptions in these analyses were that the stress-strain relationship for the rock was linearly elastic and that the inter-stope rib and chain pillars carried no significant load (ie the hangingwall was arching across the whole sandfilled area). These two assumptions were not considered to limit the validity of the empirical evaluation of a "critical" stress associated with observations of rock failure.

MINING AFTER 1974

Mining operations continued, stoping upwards from the 660m level in a V-shaped configuration. A reduction in the production rate enabled the recommended attention to be given to the mining of each stope, though the amount of support work carried out was limited while experience and confidence were gained.

Mining below the 660m level commenced, this time mining down-dip to the 730m level. No major changes were made to the mining method except to increase the thickness of the crown pillar, the stopes then being similar to those shown in Figure 10, except that the pillars were left intact and the stopes sandfilled.

Stoping continued without major problems until February 1977 when major rockbursts occurred throughout the sandfill area and especially on the 580m and 594m levels. Severe bumping and hangingwall collapse took place in the upper levels over a period of several days with additional stressing down-dip and significant surface subsidence being noted. Geotechnical investigations were again initiated, this time to evaluate possible maximum subsidence under the river, future mining conditions using sand-fill mining methods and other options for continued mining in the area.

GEOTECHNICAL ANALYSES FOLLOWING 1977 ROCKBURSTS

Analyses of the events leading upto the 1977 rockbursts were completed using MSIM3D, however the model used differed from previous models in two respects. First, the stress-strain relation-ship for the unmined ore on 580/594m levels was modified to take into account compressive rock failure that would occur as the applied stress approached the in-situ rock strength. Second, it was assumed that the mined out and sandfilled stopes above and below the 580m level could carry some fraction of the hangingwall load due to gradual failure and subsidence.

For the general form of the stress-strain relationship described earlier, assumed values for peak and residual stress are shown in Table 2. For the unmined ore, the value for peak stress was estimated from measured intact rock strength values (typically 180 MPa unconfined) and a conservative estimate of maximum stress concentration factors around orebody drives (less than 4). A range of values for residual stress was assumed for sensitivity analyses. Deformation characteristics of the sandfilled stopes (and remnant pillars) were not known, but the system was assumed to behave initially as a brittle rock (to pillar failure), then to carry a residual stress due to confinement by the sandfill. The peak stress was estimated as the load-bearing capacity of the slender (4:1 height to width ratio) pillars and was, in fact, exceeded in all of the analyses. The residual stress value was assumed to be less than the pre-mining overburden load, which ranged from 8.75 MPa to 18.25 MPa between the upper and lower mining levels. This was supported by the lack of significant surface subsidence. Zero stress values for the sandfill were used for several of the analyses, to represent an extreme case of no hangingwall subsidence (complete arching).

Twelve analyses were completed, including three to re-assess all of the rockburst events in the period 1973-1977 and to validate the model.

Table 2.   Assumed values for peak and residual stress.

	Peak Stress (MPa)	Residual Stress (MPa)
Unmined ore 580/594m level	45	0 to 36
Sandfill above 580m level	0 or 10	0 or 6
Sandfill below 580m level	0 or 13	0 or 8

For all analyses, values of E and ν were assumed to be 3.45×10^7 kPa (pre- and post-peak deformation) and 0.2, respectively.

Results for December, 1973 are shown in Figure 7. Based on these initial analyses, the correlation between rock damage and calculated "critical" stress was revised to the following guidelines.

1. Potential rockburst areas were identified where the stress exceeded 42 MPa (as expected, this is close to the assumed peak stress value).

2. Major damage (spalling) was identified where the stress was between 34 MPa and 42 MPa.

3. Minor damage (flaking) was identified where the stress was between 27 MPa and 34 MPa.

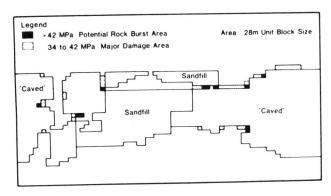

Figure 7. Predicted stress and damage criteria (MSIM3D) December 1973

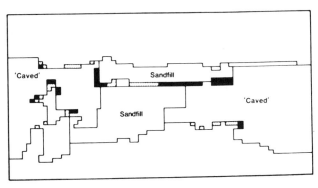

Figure 8. Predicted stress and damage criteria (MSIM3D) January 1977

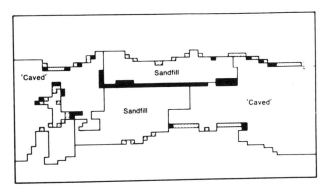

Figure 9. Predicted stress and damage criteria (MSIM3D) January 1978 (proposed schedule).

Based on the results of nine sensitivity analyses for the mining geometry as of January 1977 (Figure 8) and proposed for January 1978 (Figure 9), and a review of the potential maximum subsidence over the area, the following conclusions were drawn for future mining in this area.

1. Surface subsidence of up to 4 meters was likely, with the possibility of water from the river entering the mine through surface cracks to the caving hangingwall.

2. Mining below the 660m level would be subject to increasing levels of stress even with the down-dip mining sequence; major damage conditions were predicted on all stope faces by January 1978 with the expectation that rockburst conditions would again develop as mining continued.

3. Any further mining in the ore remnants above 660m level would increase surface subsidence in the long term by removing a major supporting pillar, but in the short term would probably transfer additional stress down-dip.
It was therefore decided to leave the 580m/594m level remnant, containing 2.8 million tonnes of ore at 3.5 per cent copper, to be mined in the future.

4. A consistently applied system of ground support should be used in development drives and crosscuts, for areas identified with high rockburst potential or major damage (greater than 34 MPa).

CHANGE TO TOTAL EXTRACTION METHODS
Because the possibility of water inflow from the stream could not be eliminated, a decision was taken to divert the river and carry it across the subsidence area in pipes. As stated earlier this option had been studied previously and so various schemes based on known hydrological data were immediately available. The basic design parameters were as follows.

1. The river would be dammed at a point beyond the 70° caving angle drawn from the base of the known ore reserves.

2. From this dam the river would flow through two 850mm diameter pipes, 900m in length and discharge over the stable footwall bedrock.

3. Flexible couplings would be used in the pipe line to allow for subsidence and subsequent

re-levelling operations.

At the same time that construction was in
in progress, further studies were initiated to
evaluate alternative mining methods that could
be used beneath this area. In support of these
studies, additional MSIM3D analyses were undertaken in 1978 to evaluate total extraction mining,
with and without regional support pillars. The
results of these analyses, combined with an
economic assessment of both methods, led to the
decision to revert to total extraction mining,
with hangingwall caving, on completion of the
diversion in 1978.

MINING METHODS

Initially the changeover to total extraction
mining methods did not require too many changes
to the design of the stope as, instead of filling
the stope with sandfill, the stope was left open
and the pillars were drilled and then blasted
into the open stope. (Figure 10)

Nonetheless problems were encountered with
this method due to a less than ideal dip of the
orebody for open stoping and the effect of the
crown pillar 'punching' through onto the footwall
drive.

This was partly solved by using the 'Cascade'
or continuous retreat open stoping method
invented at Mufulira a few years previously.[3,4]
This method was better suited to the lower dip
of the orebody and did not require a footwall
drive. (Figure 11, 12)

In the past few years the stoping method
has reverted back to open stoping because of an
increase in dip, the Cascade method only being
suitable for dips of under 35^0.
Throughout the whole of the period the area has
been under stress; the fact that mining has
successfully taken place without major mishap,
despite rock spalling and minor bursts, is
largely due to the development and use of good
support techniques.

Support Techniques

Prior to 1973 very little support was required
and where it was it generally took the form of
mechanically anchored rock bolts. In some areas
timber packs were installed and these were often
rockfilled. (Figure 13). Following the rockbursts in 1973 and 1974 (Figure 14) it was noted
that rockbolted areas had prevented some rockfalls.
This, together with the recommendations made as
a result of the geotechnical analyses, provided
the impetus to use fully grouted steel dowels.

Initially this took the form of resin grouted
deformed reinforcing bar and, although successful
from a geotechnical view point, it suffered the
disadvantage of cost, difficulty of installation
and limited shelf life of the resin packs.

A major improvement was achieved by replacing
the resin with a thick cement-water grout. The
grout is mixed on site and emplaced with a Spedel
pump, a light weight air pump. It is possible
for a 3-man crew to place 100 bolts in a shift,

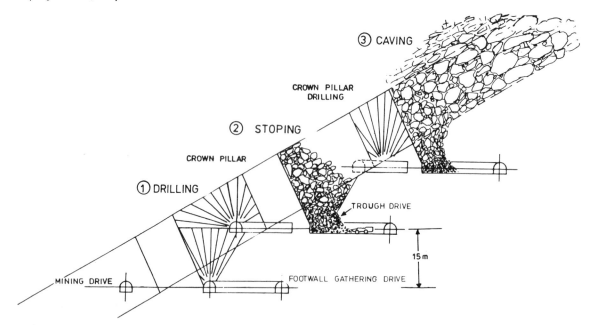

Figure 10.   Section showing mechanised open stoping method. Note position of footwall gathering
drive below crown pillar

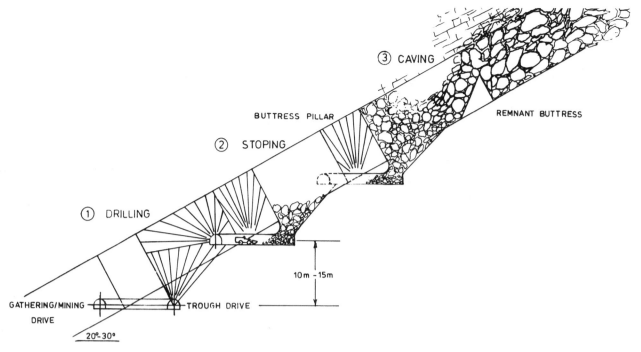

Figure 11. Section showing Cascade method

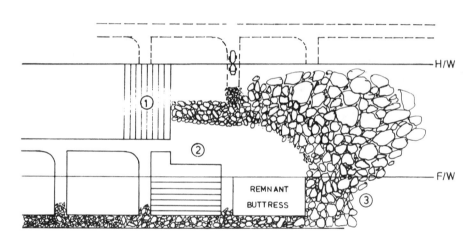

Figure 12. Composite plan showing Cascade method

Figure 13. An installation of timber packs

Figure 14. Unsupported drive showing severe spalling damage

Figure 15. Pigtail ends on steel dowels with mesh and lacing

Figure 16. Typical drive showing diamond pattern

Figure 17. Spalled rock contained by wire mesh

Figure 18. Closure crosscut with timber lagged yielding steel arch sets

this crew being responsible for mixing and pumping the grout and placing the 2.4m dowels in pre-drilled holes. The dowel material most commonly used is 16 - 20mm mild steel reinforcing bar, but on occasions due to supply problems, smooth bar has been installed. Subsequent hydraulic pull testing has shown that, as long as the grout is correctly installed, either type of steel breaks before shearing of the steel-cement bond occurs.

When available, destranded and straightened Langs-lay discard hoist rope is utilised. This has the advantage of being a cheap source of an excellent dowelling material but is slow to prepare and requires a larger hole to be drilled. The stress levels encountered have necessitated the almost universal use of wire mesh and lacing wire. To facilitate this all of the roof bolts now emplaced have either a pigtail or a shepherds crook formed on their outer end (Figure 15).

The lacing rope used is also reclaimed from destranded hoist rope and is clamped with a locally made device made from 2 deformed (figure of eight) chain links and bolts. The pattern normally used is a 1.5 by 1.0m diamond grid and the dowel length generally 2.4m, although on some intersections and other large span areas the length is increased to 3.8m (Figure 16).

Great care is taken with quality control and random hydraulic pull testing is carried out.

Due to various constraints it has never proved possible to install the support system as close behind the development as would be ideal. However, it has been proved that, even if some relaxation and deterioration has occurred, this system will normally contain and support the highly stressed rock which is encountered near to the mining faces in this area. (Figure 17).

In closure areas, especially final retreat crosscuts, experience has dictated the use of closely spaced yielding arch sets. (Figure 18). Following final closure a reasonable percentage of these steel sets can often be recovered from the footwall drives.

CONCLUSIONS

The use of geotechnical analyses and in particular the mining simulation program MSIM3D proved to be an invaluable aid to mine management in making decisions about mining the area. Certainly the magnitude and implications of the decisions were immense; the construction of a ZMK8 million dam complex to divert the river, the shortfall in production while stoping sequences were changed and the leaving of the 580m/594m level remnant containing 2.8 million tonnes of ore at 3.5 per cent copper.

Since 1978 about 6 million tonnes of ore have been mined without major problems from an area which, to this day, is still subject to stress. The fact that this has been achieved with a high degree of safety is due to the development of an effective rock support system.

Nonetheless the mining of the area would not have been possible without the skill and determination of the personnel involved.

In the future further geotechnical analyses will be carried out to assess the viability of mining the 580/594m level remnant.

ACKNOWLEDGEMENT

The authors wish to thank the management of Zambia Consolidated Copper Mines Limited for permission to publish this paper. They would also like to thank their colleagues for the assistance given in the preparation of the paper.

References

1. O'Connell C.A. The Mufulira Western expansion project increases production 50%. In: Mining Engineering, December 1962, 61-67.
2. Starfield, A.M. and Crouch, S.L. Elastic analysis of single seam extraction, In: H.R. Hardy, Jr. and R. Stefanko. (Editors) New Horizons in Rock Mechanics, A.S.C.E., New York, 1983, PP. 421-439
3. Scott F.D. Development of the Cascade method of continuous retreat open stoping at Mufulira Copper Mines Limited, Zambia. Transactions Institution of Mining and Metallurgy (Section A), 79, 1970, A96-104.
4. Airey L.D. Introduction of the Cascade method of continuous retreat open stoping at Mufulira Copper Mines Limited, Zambia. Transactions Institution of Mining and Metallurgy (Section A), 75, 1966, A137-46.